Functional Genomics

A Practical Approach

Edited by

Stephen P. Hunt

Department of Anatomy, University College London U.K.

and

Rick Livesey

Department of Genetics, Harvard Medical School, U.S.A.

OXFORD
UNIVERSITY PRESS

Great Clarendon Street, Oxford OX2 6DP

Oxford University Press is a department of the University of Oxford.
It furthers the University's objective of excellence in research,
scholarship, and education by publishing worldwide in

Oxford New York

Athens Auckland Bangkok Bogotá Buenos Aires Calcutta Cape Town
Chennai Dar es Salaam Delhi Florence Hong Kong Istanbul Karachi
Kuala Lumpur Madrid Melbourne Mexico City Mumbai Nairobi Paris
São Paulo Singapore Taipei Tokyo Toronto Warsaw

with associated companies in Berlin Ibadan

Oxford is a registered trade mark of Oxford University Press in the UK
and in certain other countries

Published in the United States by Oxford University Press Inc., New York

First published 2000

A catalogue record for this title is available from the British Library

Library of Congress Cataloguing in Publication Data
(Data available)

1 3 5 7 9 10 8 6 4 2

ISBN 0 19 963775 X (Hbk.)
ISBN 0 19 963774 1 (Pbk.)

Typeset in Swift by Footnote Graphics, Warminster, Wilts
Printed in Great Britain on acid-free paper
by The Bath Press, Avon

Preface

It has been a great pleasure to prepare this multi-author text on functional genomics techniques at such an exciting time in biology. Without wishing to add to the chorus of hyperbole surrounding the completion of the first human genome sequence, the availability of the complete genome sequences of many organisms is already having a significant impact on biomedical research. It is clear that the availability of those sequences provides a powerful resource for studying biological processes, but it is also changing the way in which many scientists carry out research. Biomedical scientists in general, and molecular biologists in particular, are starting to undertake, for want of a better term, systems-level analyses of biological processes, starting with whole-genome expression analyses. This first wave of studies is predicated on the hypothesis that knowledge of which genes are expressed in different cell types under different conditions will allow the prediction of gene expression networks (or network architecture), thereby uncovering the logic of transcriptional control. Such analyses at the transcriptional level will be accompanied by similar analyses at the protein expression level, leading to the development of an integrated model of cellular gene and protein expression dynamics. Therefore the focus of this book is on the current techniques available for gene and protein expression analysis, in order to facilitate the use of these technologies in a diverse range of settings. We hope that readers will find this a useful snapshot of the currently available technologies, a starting point for evaluating the applicability of these approaches to their own research and also a helpful benchmark for evaluating new and improved emerging technologies. Finally, we would like to express our gratitude to all of the authors who have generously given of their time and expertise to write an excellent series of chapters on current functional genomics methodologies.

Boston, September 2000 — Rick Livesey
London, September 2000 — Stephen Hunt

Contents

Protocol list

Mass spectrometry

Abbreviations

Ad1	adaptor 1
Ad2R	adaptor 2R
AZT	anilinothiazolinone
CAPS	3-[cyclohexylaamino]-1-propanesulphonic acid
cDNA RDA	cDNA representational difference analysis
CEA	carcinoembryonic antigen
CIP	calf intestinal phosphatase
DEPc	diethylpyrocarbonate
dNTP	deoxynucleotide triphosphate
DRG	dorsal root ganglion
ds	double stranded
DTT	dithiothreitol
ESI MS	electrospray ionization mass spectrometry
EST	expressed sequence tags
EtBr	ethedium bromide
FACS	fluorescence-activated cell sorter
G3PDH	glyceraldehyde 3-phosphate dehydrogenase
IPG	immobilized pH gradients
MALDI TOF	matrix-assisted laser desorption/ionization time of flight
MBN	mung bean nuclease
MDS	multidimensional scaling
MMLV	Moloney murine leukaemia virus
MOPS	4-morpholinepropane-sulphonic acid
NP1	nested primer 1
NP2R	nested primer 2R
PCD	peptide-collecting device
PCI	phenol–chloroform–isoamyl alcohol
PCR	polymerase chain reaction
PMF	peptide mass fingerprinting
PMSF	phenyl sulphonyl methylfluoride
PSA	prostate specific antigen
PTC	phenylthiocarbamyl peptide
PTH	phenylthiohydantoin

RDA	representational difference analysis
ss	single stranded
SSH	suppression subtractive hybridization
TAE	Tris-Acetate-EDTA
TBE	Tris-Borate-EDTA
TEMED	tetramethylethylenediamine
TEN	Tris-EDTA-Nacl
TFA	trifluoroacetic acid
TPEA	three prime end amplification

Chapter 1
Functional genomics: approaches and methodologies

F. J. Livesey
Department of Genetics and Howard Hughes Medical Institute, Harvard Medical School, Boston, MA, USA.

Stephen P. Hunt
Department of Anatomy and Developmental Biology, Medawar Building, University College London, UK

1 Introduction: functional genomics

We now study biological processes in the context of finite genomes. To give one example, the mammalian genome is estimated to contain 100 000 genes, of which typically 12–15 000 are transcribed in a given cell. Thus, the functional differences between cell types are ultimately due to the differences in the genes expressed in each cell, both during development and in the terminally differentiated cell. Furthermore, every process within the whole organism, from embryonic development to learning and memory to the response to infection, is dependent on the deployment of this core 'toolbox' of 100 000 genes.

Selective deployment of the genome is as important for prokaryotes with genomes of 2000 genes as it is for humans with 100 000 genes. The purpose of functional genomics is to place all of the genes in the genome within a functional framework, both in the most basic terms of what the protein encoded by each gene does and also in the broadest sense of the role of each gene in the overall functioning of the cell and organism. This definition of functional genomics carries the clear implication that studies of gene function must advance on both of these levels simultaneously.

2 Current methods in functional genomics

Functional genomics rests on a few key technical approaches: expression profiling, or mRNA and protein expression measurements; reverse genetics, the generation of targeted mutations in genes of interest; mutagenesis screens (or forward genetics), the generation of random mutations in the genome and the screening of the pool of mutants for phenotypes of interest; and bioinformatics, the analysis of data generated by all of the above approaches, but particularly expression profiling, in order to extract the maximum information.

The combination of these approaches is proving to be a very powerful tool for understanding biological processes at the molecular level. For example, such combinations can be used to identify key gene expression networks that are important in the development of a particular structure. In this case, one can identify components of a key gene expression pathway or network by comparing gene expression in a developing structure under normal conditions with gene expression in that developing structure in animals carrying targeted mutations in a gene known to be important for the development of that structure.

By assembling a number of such datasets from animals carrying mutations in different genes involved in its development, one can reconstruct many of the gene expression pathways regulated by these genes, predict where they interact, and identify novel genes regulating this process. In the case of mammalian development, there is no shortage of animals carrying targeted mutations in developmentally important genes that can be used for such studies. Such approaches will also be very useful for understanding the genetic defects in mutant animals generated by mutagenesis screens, particularly in organisms, like mouse and zebrafish, where it is not possible to tag mutation sites as they are generated.

One fundamental approach to understanding gene function is to study where and when each is expressed. There have been many studies carried out over the last two decades that have attempted to identify genes preferentially expressed in particular tissues or cell types. However, it is only in the last few years that concentrated efforts have been made to develop and implement methods to systematically study differences in gene expression at the whole-genome level. One important stimulus to this effort has been the progress of the international genome sequencing projects and their soon-to-be-realized goal of identifying every gene in the genome of several model organisms. As this book goes to press, yeast and *C. elegans* have been completed, the first pass human sequence is not far away, and the *Drosophila* and mouse genomes have a good deal left to complete. Therefore, we will soon know the sequence of every possible transcript (and its corresponding protein) from the genomes of several organisms, but we will have no understanding of the functions of these transcripts and their corresponding proteins.

As the range of chapters in this book testifies, we now have the technologies to identify all of the genes and proteins expressed in a given cell type or tissue. The assumption we make is that, by observing what is going on at the level of transcription and translation, we can deduce a great deal about the biological process under study. The challenge is to make sense of all of this gene and protein expression data in order to reconstruct genetic control networks, to assign possible functions to novel genes and proteins, and to understand biological processes at the molecular level.

However, this challenge demands that an understanding of the dynamic control of gene expression be integrated with our increasing understanding of signal transduction, cell signalling and overall cell function. The first step in this is to understand the dynamic regulation of gene expression in any given cell type. Large gene and protein expression datasets will allow researchers to uncover patterns within and across those data. It has been proposed that such patterns will identify genetic regulatory networks, or groups of genes that are regulated in similar ways and are involved in similar biological processes (1, 2). Several studies testing this principle have been carried out recently, both in yeast and in human cell lines (3–5).

These studies have demonstrated the usefulness of applying exploratory methods such as cluster analysis for finding patterns in datasets. For example, in both yeast and human cells, transcripts associated with different parts of the cell cycle form discrete clusters when gene expression is studied over a time course in cells synchronized for the cell cycle (3–5). In addition, these studies allowed expressed sequence tags encoding proteins of unknown function to be assigned to putative functional classes, based on their clustering with genes of known function. The next step in functional genomics will be to test those putative functions and apply these methods to complex biological processes.

3 About this book

Within this book, we have chosen to emphasize the area of expression profiling, as it is the component of functional genomics that is most widely used in research laboratories. In addition to its usefulness in its own right, expression profiling allied with the proven uses of forward and reverse genetics and the power of bioinformatics is becoming a powerful approach for studying the architecture of genetic networks and the reconstruction of the systems underlying cellular functions.

The aim of this book is to gather a number of different technical approaches in one volume with detailed, practical descriptions of these methods by their inventors and experienced users. In all cases, the aim has been to supply enough practical information and detailed protocols to allow scientists to judge the usefulness and applicability of each method to the biological questions that interest them and to implement those protocols in their own laboratory. The majority of the techniques describe different technical approaches to identifying genes that are differentially expressed between tissues. In many cases, however, it is desirable to know as much as possible about the entire complement of genes expressed in a given cell or tissue. In such cases, absolute gene expression information is what is required. One approach to characterizing every RNA transcript in a given tissue is serial analysis of gene expression, or SAGE (6), and Chapter 7 describes a modified SAGE protocol for small amounts of tissue, termed SADE.

It should be noted that RNA levels in themselves do not appear to be reliable indicators of intracellular protein levels. Therefore, we have also included two chapters on the principles and practice of proteomics, or protein expression profiling. As discussed in both chapters, current methods for identifying proteins and peptides in complex mixtures are reaching very high levels of accuracy and sensitivity, allowing for expression profiling of the more abundant proteins and comparison of protein expression between tissues.

Investigators must typically weigh a number of factors when deciding between technical approaches for gene expression studies. The main issues include the sensitivity of each system, their ease of use, the amount of biological material needed, overall cost, and infrastructure requirements. It is for this reason that this book contains a diverse range of techniques—all have been shown to work efficiently and are currently in use in laboratories around the world, but each differs in its technical requirements.

The most important consideration for investigators considering applying functional genomics methods to their area of interest is the technical requirements of each method. The methods described here cover a range in technical difficulty and require varying levels of expertise. Several of the methods described in this book are available as commercial kits. The advantage of these kits is that, as the components have been performance tested and are in general use, there is little set-up or optimization time involved. In addition, the commercial kits are often simple to use, in comparison with producing and testing all of the reagents in-house. However, time invested in setting up any of these methods, testing, and optimizing them, is obviously a useful investment for laboratories that intend carrying out a number of gene expression profiling projects.

In addition, some of the methods require considerable investments in reagents and infrastructure. At current costs, the investment in setting up a microarray production and analysis facility is well beyond the budget of many single laboratories, although commercial systems are driving the costs down. This is in marked contrast to the cost of buying any of the commercial kits available for many of the techniques described here. Such considerations are particularly relevant for investigators whose primary interest is not in functional genomics, but would like to carry out some genomics studies to broaden their attack on a particular problem.

Regardless of cost and infrastructure requirements, however, it is logical to ask whether any of these approaches is clearly more efficient at identifying gene expression differences than the others. This is a particularly appropriate question regarding the differential gene expression methods, as both SAGE and two-dimensional gel electrophoresis of proteins are very efficient methods for absolute transcriptional profiling. Both give information on RNA or protein abundance, as well as identification of genes and proteins expressed in a given tissue. This is in contrast with the differential expression methods that are designed to identify transcripts that are enriched in one or more of the samples studied.

All of the methods described here, subtractive hybridization, differential display, representational difference analysis (RDA), suppression subtraction hybridisation (SSH) and cDNA microarrays, have been successfully used by a number of different laboratories. With the exception of microarrays, none require investment in new machinery, and all use standard molecular biological techniques. Differential display, SSH, and RDA are all very useful where tissue is limiting and little RNA is available. The advantage of subtractive hybridization, however, is that one has a stock of an enriched library that contains cDNA inserts of a good size for gene identification and can be analysed exhaustively over time.

The newest of the approaches described, and one generating a great deal of interest, is cDNA microarrays. The advantage of microarrays is that one can study gene expression at the whole genome level, something which is theoretically possible with the other methods, but difficult to be certain of. In addition, from the microarray data published to date, it appears that microarrays are sensitive, accurate, and do not have a problem with false positive results, a common problem in differential gene expression profiling. One of the most exciting features of microarrays is the ability to analyse datasets from a number of experiments to identify genes of particular interest and patterns in gene expression. However, microarray production requires considerable investment in infrastructure and materials, although there will be commercially produced microarrays available in the coming years.

In conclusion, deciding which method is best suited to a particular problem or laboratory is a balance between scientific issues, such as the amount of RNA or protein available and the technical experience of the laboratory, and general issues, such as equipment availability and the projected use of functional genomics methods by the laboratory over time.

This book is a sampling of the range of techniques available to researchers in the field of functional genomics. As with all samplings, it has its omissions and, no doubt, its inadvertent biases. In the case of omissions, we have not included chapters on oligonucleotide arrays for transcriptional profiling, as typified by the Affymetrix GeneChips, or on mutagenesis or genetic screens in any organism. Oligonucleotide chips are extremely useful tools for transcriptional profiling, resequencing, and sequence polymorphism detection. However, from the point of view of transcriptional profiling, cDNA microarrays are becoming a more widely available resource for biomedical scientists. Although we would have liked to include a piece on mutagenesis screens, we felt that this topic could not be dealt with in sufficient depth in a single chapter and would require another volume entirely to cover the range of systems and organisms available.

One other obvious deficiency is in the area of bioinformatics, a rapidly evolving field in its own right. As emphasized above, of particular relevance to functional genomics is the area of data mining, or extracting maximum information from gene and protein expression data in order to reconstruct the systems operating within cells. While we would have liked to include a chapter on this topic, given the rapid evolution of this field, it is likely that anything included would have been out of date before the book was in print.

4 Future developments in functional genomics

Given the speed with which functional genomics has advanced in the last decade, one could ask where we might go from here and whether there are potential developments around the corner that will make much of the current technology redundant. There are some more advanced approaches currently being developed in a number of laboratories, in particular in the area of parallel sequencing; however, most of the next wave of developments will be in the application of current techniques to new problems or their adaptation to new uses.

One growing application of gene expression profiling is in understanding human pathology. Studies are underway of the gene expression response of human cells and tissues to disease, of hosts to infectious disease and of how pathogens modify gene expression during infection. It is hoped that such studies will yield fundamental insights into the aetiology and pathogenesis of human disease. Such a proposal is dependent on the efficient management and mining of the expression data generated, such that key pathways or interactions can be identified as important for disease causation and progression and as candidates for therapeutic intervention.

As more and more differential gene and protein expression studies are carried out, one of the most pressing issues in functional genomics is the integration of data generated in different laboratories using different approaches. Unfortunately, integrating very different datasets is not as straightforward as assembling a sequence database. In the case of sequence databases, there is a standard end-point of the sequence itself, which serves as an absolute standard. Although standardized databases could be constructed for transcriptional profiling methods based on absolute RNA and protein levels, this is clearly not the case for relative gene expression data (7).

A general problem with all expression profiling methods is their need for considerable amounts of RNA or protein, which in turn requires a good deal of tissue. A number of methods have been developed for amplifying small amounts of cDNA, using the polymerase chain reaction (**PCR**) (8–10) and T7 RNA polymerase (11). Several of these methods have been successfully used to make and amplify cDNA from sources as limited as thousands of cells for further characterization by expression profiling (12), and as little as a single cell for PCR and Southern blot analysis (8, 9, 11), as described in Chapter 8. The optimization of these methods will allow the application of expression profiling methods to, for example, clinical biopsy samples and the early stages of vertebrate development.

These limitations are even more severe for protein expression analysis, as there are no methods currently available for protein amplification. Therefore, in order to increase the number of proteins that can be analysed from a single sample and to decrease the amount of tissue needed, there is a great deal of interest in developing more sensitive protein detection and identification methods.

There have been optimistic predictions made about the potential of 'post-genomic' research. There is an enormous potential to be realized from the study of genomes in action, and it can revolutionize our understanding of biology. This potential can only be realized by biologists working alongside scientists from other disciplines, particularly mathematics, chemistry, and engineering to apply all of this potential to the study of biological questions. We hope that the excellent accounts gathered in this volume will introduce biological and biomedical scientists to the potential uses of functional genomics in their work and encourage them to apply these methods where appropriate.

References

1. Wen, X., Fuhrman, S., Michaels, G. S., Carr, D. B., Smith, S., Barker, J. L., and Somogyi, R. (1998). *Proc. Natl. Acad. Sci.*, **95**, 334–339.
2. Eisen, M. B., Spellman, P. T., Brown, P. O., and Botstein, D. (1998). *Proc. Natl. Acad. Sci.*, **95**, 14863–14868.
3. Iyer, V. R., Eisen, M. B., Ross, D. T., Schuler, G., Moore, T., Lee, J. C.F., Trent, J. M., Staudt, L. M., Hudson Jr., J., Boguski, M., Lashkari, D., Shalon, D., Botstein, D., and Brown, P. O. (1999). *Science*, **283**, 83–87.
4. Tavazoie, S., Hughes, J. D., Campbell, M. J., Cho, R. J., Church, G. M. (1999). *Nat. Genet.*, **22**, 281–285
5. Spellman, P. T., Sherlock, G., Zhang, M. Q., Iyer, V. R., Anders, K., Eisen, M. B., Brown, P. O., Botstein, D., Futcher, B. (1998). *Mol. Biol. Cell*, **9**, 3273–3297
6. Velculescu, V. E., Zhang, L., Vogelstein, B., Kinzler, K. W. (1995). *Science*, **270**, 484–487
7. Ermolaeva, O., Rastogi, M., Pruitt, K. D., Schuler, G. D., Bittner, M. L., Chen, Y., Simon, R., Meltzer, P., Trent, J. M., Boguski, M. S. (1998). *Nat. Genet.*, **20**, 19–23
8. Brady, G., Billia, F., Knox, J., Hoang, T., Kirsch, I. R., Voura, E. B., Hawley, R. G., Cumming, R., Buchwald, M., Siminovitch, K., Miyamotot, N., Boehmelt, G., and Iscove, N. N. (1995). *Current Biology*, **5**, 909–922.
9. Dixon, A. K., Richardson, P. J., Lee, K., Carter, N. P., Freeman, T. C. (1998). *Nucleic Acids Res.*, **26**, 4426–4431
10. Chenchik, A., Zhu, Y. Y., Diatchenko, L., Li, R., Hill, J., and Siebert, P. D. (1998). In *Gene Cloning, and Analysis by RT-PCR*. pp. 305–319. BioTechniques Books, Natick, MA.
11. Eberwine, J., Yeh, H., Miyashiro, K., Cao, Y., Nair, S., Finnell, R., Zettel, M., and Coleman, P. (1992). Analysis of gene expression in single live neurons. *Proc. Natl. Acad. Sci. USA*, **89**, 3010–3014.
12. Luo, L., Salunga, R. C., Guo, H., Bittner, A., Joy, K. C., Galindo, J. E., Xiao, H., Rogers, K. E., Wan, J. S., Jackson, M. R., and Erlander, M. G. (1999). *Nature Med.*, **5**, 117–122.

Chapter 2
Construction and screening of a subtractive cDNA library

Armen N. Akopian and John N. Wood
Department of Biology, University College London, UK

1 Introduction

Subtractive library production and screening are powerful methods to isolate transcripts that are exclusively or selectively expressed either in certain tissues, or during development or in disease states. This chapter describes procedures for the isolation of specific transcripts using construction and screening of subtractive libraries, and illustrates examples of typical problems arising during the experiments. The original method was developed by Davis and colleagues in classical work on the isolation of T-lymphocyte-specific transcripts, including a third type of murine T-cell receptor gene (1, 2). Over the years this approach has been modified. However, the principle behind the method is essentially the same, and only a few modifications have improved on the remarkable sensitivity and efficiency of the original method. Construction and screening of a subtractive cDNA library can be divided into several steps (*Figure 1*).

(1) Preparation of nucleic acid probes for subtractive hybridization.

(2) Subtractive hybridization between probes.

(3) Cloning of the remaining cDNA to construct a subtractive cDNA library.

(4) Preparation and labelling of probes for screening of the subtractive cDNA library.

(5) Analysis of clones.

For many of these steps, a number of options exist, and the choice of procedures should meet the following conditions:

(1) Sensitivity. It should be possible to isolate cDNA clones representing mRNAs with abundance as low as 0.0001–0.001%.

(2) Efficiency. The method should require modest amounts of starting RNA and allow the isolation of as many specific transcripts as possible of wide size range.

(3) Convenience. The procedures should not be too cumbersome, and should avoid the isolation of any artefactual clones.

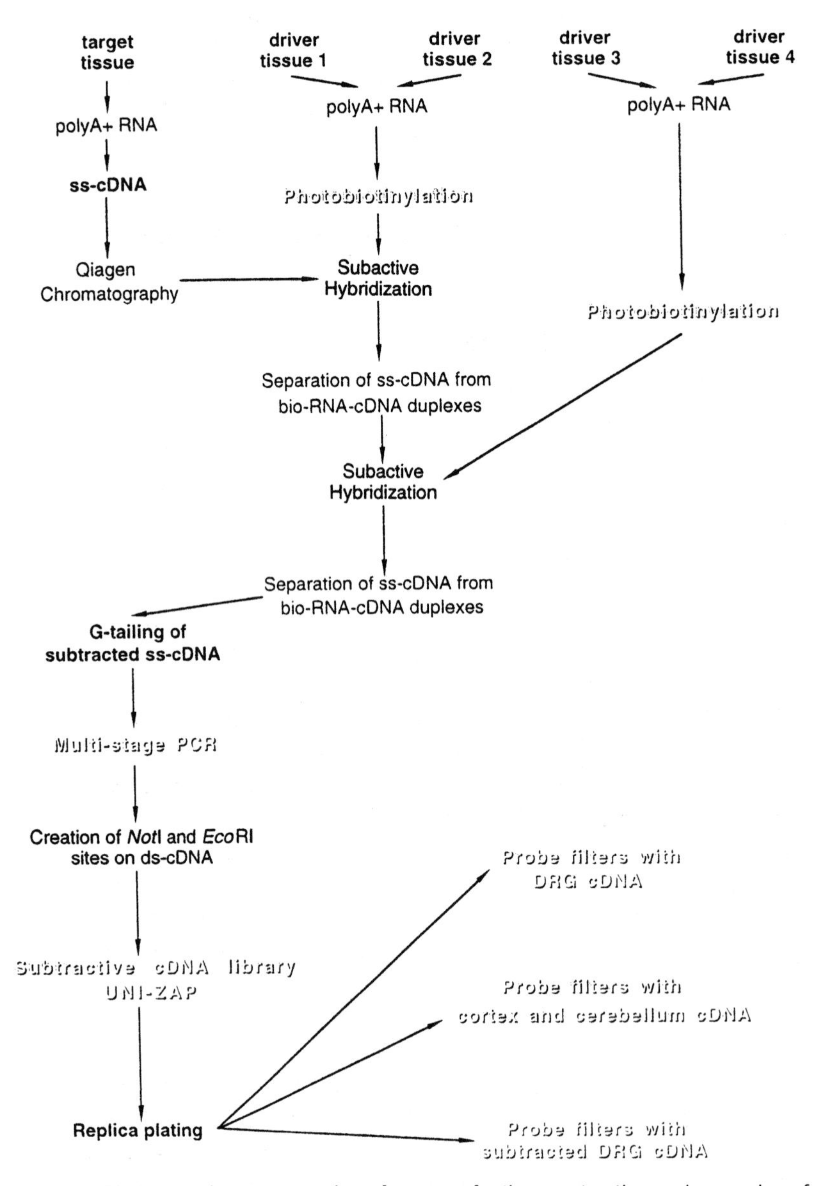

Figure 1 Diagrammatic representation of strategy for the construction and screening of subtractive cDNA library.

2 Reagents, equipment, deoxyribonuclease-free and ribonuclease-free conditions

Use molecular biological-grade reagents and ultrapure chemicals. Sources of reagents and chemicals are indicated in brackets; specialized equipment is listed in the Appendix for this chapter. Stock solution are prepared in ultrapure water obtained from a water purifier (UHQ Elgastat), and then filtered through a 0.22-

mm filter (Sartorius). Stock solutions and water are DEPC-treated and autoclaved to destroy deoxyribonuclease (DNase) and ribonuclease (RNase). Sterile ultrapure water and solutions, as well as RNase- and DNase-free disposable plastic tubes and tips are always employed during experiments.

3 Preparation of nucleic acid probes for subtractive hybridization

3.1 Type of probes for target and driver

Subtractive hybridization typically involves annealing of nucleic acids (poly(A)$^+$ RNA or single-strand cDNA (ss-cDNA) or double-strand cDNA (ds-cDNA)) extracted or/and synthesized from *driver* tissues with at least 20-fold less ss- or ds-cDNA synthesized from another source (*target*) for which specific transcripts have to be isolated. Which type of probes for driver and target are best? To achieve a maximum of efficiency and sensitivity, contamination of target cDNA by driver nucleic acids after hybridization has to be completely avoided. Furthermore, one round of subtractive hybridization should result in as much specific subtraction as possible of common abundant molecules. Whatever the separation method of target cDNA from both annealed and free driver nucleic acids which coexist after subtractive hybridization, slight cross-contamination is always found. Therefore, either driver nucleic acids need to be eliminated after hybridization while target cDNA remains intact, or unseparated annealed and free driver nucleic acids must not interfere with cloning of the subtracted target sequences.

Although all kinds of combination of driver nucleic acids and target cDNA have been used by different groups (3–6), it has been observed that the harnessing of poly(A)$^+$ RNA as driver and ss-cDNA as target is the best choice for the following reasons. First, unseparated driver RNA is totally eliminated by alkaline hydrolysis, whereas target ss-cDNA withstands this sort of treatment. Secondly, several different methods of cloning of ss-cDNA can be chosen whereby free driver RNA and annealed RNA, which is a part of RNA–cDNA duplex, does not interfere in the process of cloning. Thirdly, hybridization of ss-cDNA with complementary poly(A)$^+$ RNA results in a hundred-fold specific subtraction, while hybridization between driver and target ds-cDNAs, or between target ds-cDNA and driver poly(A)$^+$ RNA does not result in such efficiency. Fourth, synthesis of target and driver ds-cDNA or driver ss-cDNA, as well as the construction of cDNA library for driver and/or target prior to subtractive hybridization, require additional difficult manipulations. Preparation of target ss-cDNA and driver poly(A)$^+$ RNA for subtractive hybridization will be described in the following paragraph.

3.2 The choice of target and driver tissues

One of a number of deciding factors in the generation of a representative subtractive cDNA library, which in turn determines the efficiency and sensitivity of the experiment, is the choice of target and driver tissues. Both the spectrum of transcripts and their abundance in the library are influenced by the target cDNA and driver nucleic acids for the reasons detailed below.

First, target tissues do not always comprise a homogenous population of cells. Therefore, it may be necessary to select the part of the target tissue which contains desired specific transcripts. For example, if dorsal somite-specific transcripts are needed, then dorsal somites have to be separated from ventral somites.

Secondly, the relative proportion of various target-specific transcripts is determined by the presence of common transcripts in target and driver tissues. Thus, the capsaicin receptor (VR1) and transcription factor Brn3c were not isolated from a dorsal root ganglion (DRG) subtractive cDNA library when cerebellum and kidney had been used as drivers, because a splice variant of VR1 is present in kidney and a splice variant of Brn3c exists in cerebellum (6, 8, 9). If skeletal muscle had been included as a driver in place of used drivers (cerebellum, cortex, liver, kidney), then a different repertoire of DRG-specific genes would be identified (6).

3.3 Preparation of poly(A)$^+$ RNA from driver and target tissues

Target and driver poly(A)$^+$ RNA can be purified from either total RNA or cytoplasmic RNA, or RNA extracted from ribosomes, which are associated with the endoplasmic reticulum. The majority of workers use total RNA to purify poly(A)$^+$ RNA (see *Protocols 1* and *2*). However, if the main objective is to isolate only specific transcripts coding channels and membrane-bound proteins, the use of target RNA extracted from membrane-associated ribosomes is desirable.

Protocol 1

Isolation of total RNA from target and driver tissues

Equipment and reagents

- RNA extraction buffer (25 g guanidine thiocyanate salt (Sigma), 1.76 ml 0.75 M Na citrate (BDH), pH 7.0, 0.75 ml 36% sodium laurylsarcosinate (Sigma), 31.2 ml water)
- Water-saturated phenol—phenol (Gibco BRL), melted for 1 h at 90 °C, was mixed with an equal volume of water, and allowed to stand for 1–2 days at 4 °C to separate organic and aqueous phases. Discard the water phase and keep water-saturated phenol at 4 °C
- Sterile 1.5 ml microfuge, 15 ml Falcon, and 50 ml Falcon polypropylene tubes
- *Escherichia coli* tRNA—1 ml of 10 mg ml^{-1} *E. coli* tRNA (Boehringer Mannheim) was treated with 50 U of DNase I (5 U μl^{-1}) (Boehringer Mannheim). tRNA was extracted twice with water-saturated phenol. After extraction, tRNA was precipitated, washed, dried, and dissolved in 900 ml water

Method

1 Disrupt 1 g of the tissue in liquid nitrogen to such a degree that it becomes a powder.

Protocol 1 continued

2 Add the disrupted tissue to RNA extraction buffer (10–20 ml g^{-1} of tissue), homogenize this mix, and than leave it for 10 min at 60 °C.[a]

3 Add 1/10 volume of 2 M sodium acetate, pH 4.0, mix and add equal volume of water-saturated phenol. Mix again and add × volume of chloroform.[b]

4 Mix vigorously for 1–2 min at room temperature and chill on ice for at least 15 min.

5 Spin at 10 000 g for 30 min at 4 °C.

6 Precipitate the aqueous phase with an equal volume of isopropanol. Incubate for at least 1 h at −20 °C, and repeat step 5.

7 Dissolve the pellet in 10 ml of a RNA extraction buffer and 1 ml of 2 M sodium acetate, pH 4.0. Add an equal volume of isopropanol, incubate for at least 1 h at −20 °C, and repeat step 5.

8 Wash the pellet using 70% ethanol, air-dry, and dissolve in water at concentration 1–3 mg ml^{-1}. Keep RNA at −70 °C (RNA may be kept at −20 °C for up to 3 months).

[a] If RNA is extracted from either a few cells, or small parts of tissues, step 1 can be omitted, and tissue can directly be homogenized in 0.5–1 ml of a RNA extraction buffer using a syringe.

[b] If we expect to extract less than 2–3 mg of total RNA, then 5–10 mg of *E. coli* tRNA have to be added.

Protocol 2

Purification of poly(A)$^+$ from target and driver total RNA

Equipment and reagents

- Sterile 1.5 ml microfuge tubes
- 2 × elution buffer (50 mM Tris, pH 7.0, 5 mM ethylenediamine tetraacetic acid (EDTA), pH 8.0, 0.2% sodium dodecyl sulphate (SDS))
- Washing solution (0.2 M NaCl, 50 mM Tris, pH 7.0, 5 mM EDTA, pH 8.0, 0.2% SDS)
- Loading solution (0.5 M NaCl, 50 mM Tris, pH 7.0, 5 mM EDTA, pH 8.0, 0.2% SDS)
- 10 × TM (500 mM Tris, pH 7.5, 70 mM $MgCl_2$)

Method

1 Add 25–50 mg oligo(dT) cellulose (Type III, Pharmacia) to 1 ml water in a 1.5 ml microfuge tube. Mix, spin for 20 s at 1500 *g*, and discard water with non-precipitated cellulose particles.

2 Add 1 ml loading solution to equilibrate the oligo(dT) cellulose, which is now ready to use.

3 Add an equal volume of 2 × elution buffer to total RNA dissolved in water (1–3 mg ml^{-1})[a] (see Protocol 1). Heat this mix for 5 min at 65 °C, quickly chill on ice and add 1/10 volume of 5 M NaCl.

Protocol 2 continued

4 Spin oligo(dT) cellulose equilibrated with loading solution (see step 2) for 20 s at 1500 g, discard the loading solution and replace with mix of total RNA (see step 3). Gently shake total RNA with oligo(dT) cellulose for 20–30 min at room temperature.
5 Spin oligo(dT) cellulose for 20 s at 1500 g, discard total RNA (it is already poly(A)$^-$ RNA) and wash oligo(dT) cellulose bearing poly(A)$^+$ RNA twice with 1 ml loading solution for 2–3 min at room temperature.
6 Wash oligo(dT) cellulose twice with washing solution for 2–3 min at room temperature.
7 Add 150 μl elution buffer to oligo(dT) cellulose, and incubate for 5 min at 50 °C. Collect elution buffer containing poly(A)$^+$ RNA using centrifugation.
8 Repeat step 7 twice more. Extract approximately 450 μl of poly(A)$^+$ RNA with equal volume of chloroform to purify RNA solution from remaining cellulose particles.
9 Precipitate poly(A)$^+$ RNA adding 1/20 volume of 3 M sodium acetate, pH 5.0, and 2.5 volume of ethanol.[b] Keep poly(A)$^+$ RNA under ethanol at least for 1 day at −70 °C.
10 Spin at 15 000g for 1 h at 4 °C, wash poly(A)$^+$ RNA pellet twice with 70% ethanol, dry, and dissolve in 10 ml water.
11 Add 8 μl water, 2 μl 10 × TM and 1 μl DNase I (5 U μl^{-1}). Incubate for 30 min at 37 °C.
12 Add 15 μl 2 M sodium acetate, pH 4.0, 115 μl water, and extract the DNase-treated poly(A)$^+$ RNA with equal volume of water-saturated phenol (see Protocol 1).
13 Precipitate poly(A)$^+$ RNA with 2.5 volumes ethanol. Keep poly(A)$^+$ RNA under ethanol at least for 1 day at −70 °C.
14 Repeat step 10. Dissolve driver poly(A)$^+$ RNA at 1–3 mg ml^{-1}, and target poly(A)$^+$ RNA in 10 μl (<1 mg ml^{-1}). Keep RNA at −70 °C.

[a] The total volume of RNA in the loading solution should be between 0.5 and 1 ml.

[b] If possible, avoid using *E. coli* tRNA to precipitate driver poly(A)$^+$ RNA, as the presence of tRNA in the subtractive hybridization solution may reduce an efficiency of subtraction. If less than 100–500 ng of driver poly(A)$^+$ RNA is expected, co-precipitation of poly(A)$^+$ RNA with 1 μg tRNA is advised. However, co-precipitation of target poly(A)$^+$ RNA with tRNA is harmless, because target ss-cDNA will be purified from free adapter-primers and tRNA (see Protocol 3).

A critical aspect of poly(A)$^+$ RNA isolation is that RNAs from both target and driver tissues have to be completely DNA-free. Even a minor contamination (few nanograms) of RNA by genomic DNA will lead to the generation of a great number of artefactual clones in a subtractive cDNA library, which contains inserts of genomic DNA. Total RNA always contains the remains of genomic DNA, regardless of what method have been used to extract RNA. Affinity chromatography on oligo(dT) cellulose significantly reduces contamination of RNA by genomic DNA. None the less, treatment of poly(A)$^+$ RNA with DNAse I to eliminate genomic DNA completely is a critical requirement (see *Protocol 2*).

Alternative methods of total RNA extraction, such as the guanidine-thiocyanate/caesium ultracentrifugation method may be used (7).

3.4 Synthesis of target ss-cDNA with adapter-primer

Two choices need to be made for the preparation of target ss-cDNA: the type of primer to be used for ss-cDNA synthesis and the method of removing primers from the cDNA. A number of different primers—oligo(dT), adapter-primer, random-primer—can be used to synthesize ss-cDNA. The efficiency of subtractive hybridization of ss-cDNA with complementary poly(A)$^+$ RNA is quite high, irrespective of whether cDNA is synthesized with random primers or oligo(dT). However, random primer-synthesized cDNA is relatively difficult to clone, because ligation of ds-cDNA with adapters and attendant processes, like phosphorylation of cDNA and separating cDNA from free adapters, have to be carried out. Moreover, an entire coding region is more likely to be cloned from an oligo(dT)-primed cDNA subtractive library rather than from a random-primed cDNA library. A further advantage of oligo(dT)-primed ss-cDNA is that the heterogeneous cDNA pool can be amplified by the polymerase chain reaction (PCR) using non-specific primers, such as oligo(dT) and oligo(dC) (see Section 4). To simplify the cloning of target ss-cDNA after subtractive hybridization, the use of adapter-primers is preferable to oligo(dT). The best plan to be followed is to include sequences of oligo(dT) with a *Not*I site in an adapter-primer (see *Protocol* 3). First, such adapter-primers anneal to the poly(A)-tail of mRNA. Secondly, *Not*I sites are a rarity in cDNA sequences. Therefore, the resulting subtracted target ds-cDNA is unlikely to be cut with *Not*I during the process of generating *Not*I sites for ligation into appropriate vectors.

ss-cDNA is produced after subtractive hybridization. To convert ss-cDNA into ds-cDNA, either non-specific primers (like oligo(dG) or oligo(dT)) should be tailed to the 5′-end of ss-cDNA, or some specific primer should be ligated to ss-cDNA. The presence of free adapter-primers together with the ss-cDNA after subtractive hybridization will result in tailed or ligated specific primers. Thereafter these extended adapter-primers with non-specific or specific primers will be converted into short ds-cDNA, ligated into vectors, and give rise to a number of artificial clones in the subtractive cDNA library. Moreover, if polymerase chain reaction (**PCR**) is used to amplify the ss-cDNA (see Section 4), 10^{-4} mg of free adapter-primers can generate 80–90% artificial chimaeric clones in the subtractive cDNA library. Therefore, it is critically important to remove all free adapter-primers prior to subtractive hybridization (see *Protocol 3*).

Figure 2 Unsubtracted (Uns cDNA) and subtracted (Sub cDNA) DRG ss-cDNA (see *Protocols 3* and *5*). cDNA was denatured by 50 mM NaOH for 5 min at 37 °C, and loaded on 1% agarose gel soaked overnight in 30 mM NaOH, 1 mM EDTA. Gel was run using 30 mM NaOH, 1 mM EDTA buffer. Gel was dried and autoradiographed for 2 h at room temperature.

Protocol 3

Synthesis and purification of target ss-cDNA

Equipment and reagents

- STE (100 mM NaCl, 20 mM Tris, pH 7.5, 1 mM EDTA, pH 8.0)
- Phenol is melted for 1 h at 90 °C, mixed with an equal volume of Tris-base, pH 12–14, and allowed to stand at 4 °C overnight to separate organic and aqueous phases. The aqueous phase is discarded. Keep Tris-saturated phenol, pH 7.0–8.0, at 4 °C
- Qia 0.5 solution (50 mM (MOPS), pH 7.0, 1 mM EDTA, 15% ethanol, 0.5 M NaCl)
- Qia 1 solution (50 mM MOPS, pH 7.0, 1 mM EDTA, 15% ethanol, 1 M NaCl)
- Qia 2 solution (50 mM MOPS, pH 7.6, 1 mM EDTA, 15% ethanol, 2 M NaCl)
- T-primer, sequence 5′-GCAGAGAGCGGCCGC(T)16-3′

Method

1. Heat 10 μl (0.001–1 μg) of target poly(A)$^+$ RNA (see *Protocol 2*, step 14) in 1.5 ml microfuge tube for 5 min at 70 °C. Quickly chill on ice.
2. Add at room temperature: 6 μl of 5 × reverse transcription (RT) buffer (Gibco BRL), 3 μl of 0.1 M dithiothreitol (DTT) (Gibco BRL), 1.5 μl of dNTP solution (Pharmacia) containing 10 mM of each deoxynucleotide triphosphate, 2 μl of T-primer (1 mg ml^{-1}), 7 μl of water to 9.5 μl of poly(A)$^+$ RNA (see step 1).[a]
3. Add 0.5 μl of RNasin (40 U μl^{-1}) (Promega) and 1.5 μl of SuperScript reverse transcriptase (200 U μl^{-1}) (Gibco BRL) to the main and tracer reaction mixtures. Incubate for 1.5 h at 39 °C.
4. Mix the main and tracer reaction mixtures (60 μl). Add 140 μl of water and 4 μl of 5 N NaOH. Mix and incubate for 20 min at 70 °C.
5. Pre-equilibrate Qiagen Tip-5 column (Qiagen) with 1 ml of Qia 1 solution containing 50 μg of *E. coli* tRNA. Wash the column with 2 ml of Qia 2 solution, then with 1 ml of Qia 0.5 solution.
6. Add 800 μl of Qia 0.5 solution and 200 μl of Qia 1 solution to 200 μl of ss-cDNA (see step 4).
7. Load the above mixture on to the pre-equilibrated column. Repeat this step three times.
8. Wash the column with 10 ml of Qia 1 solution.
9. Elute the target ss-cDNA from the column with 0.6 ml of Qia 2 solution[b,c] (*Figure 2*).

[a] To measure the subtraction level, ss-cDNA synthesized from 1/20 (0.5 μl) poly(A)$^+$ RNA is labelled with 50 μCi of [^{32}P]dCTP (Amersham). A tracer reaction includes 0.5 μl poly(A)$^+$ RNA (see step 1), 5 μl of [^{32}P]dCTP (10 mCi ml^{-1}; 3000 Ci mmol^{-1}) and 11 μl of water (*Figure 2*).

[b] Target ss-cDNA will co-precipitate with biotinylated driver poly(A)$^+$ RNA.

[c] It may be expected that 1 μg target poly(A)$^+$ RNA is converted into 0.15 μg target ss-cDNA of which 1/20 is labelled.

3.5 Biotinylation of driver-poly(A)$^+$ RNA

There are two different approaches to separating target ss-cDNA and duplex target ss-cDNA with driver nucleic acids.

(1) Chromatography on a hydroxyapatite column provides a way to separate ss-DNA and/or RNA from ds-DNA or/and RNA-DNA duplex (1, 3).

(2) Biotinylated driver nucleic acids as well as target cDNA-biotinylated driver nucleic acid duplex can easily be removed by phenol–chloroform extraction when complexed with streptavidin (10).

The use of biotinylated driver nucleic acids offers several advantages.

(1) Chromatography on a hydroxyapatite column is much more difficult to perform than the removal of driver nucleic acids with the streptavidin/ phenol–chloroform method (10).

(2) Chromatography on a hydroxyapatite column is not highly effective, and 30–50% ssDNA or RNA may irreversibly be attached to the hydroxyapatite.

(3) Hydroxyapatite does not effectively separate ss-DNA from ds-DNA or RNA-DNA duplex, and about 1–2% of ds-DNA or RNA-DNA may contaminate single stranded nucleic acids.

Altogether, the employment of biotinylated driver nucleic acids to separate target ss-cDNA effectively from driver nucleic acids is favoured over the use of chromatography on a hydroxyapatite column. We have previously mentioned that poly(A)$^+$ RNA as a driver is superior to the application of ds-cDNA or ss-cDNA as driver. Therefore, a simple method for photobiotinylation is described here (see *Protocol 4*).

Protocol 4

Biotinylation of driver poly(A)$^+$ RNA

1 Mix 15 μg of driver poly(A)$^+$ RNA[a] (see *Protocol 2*, step 14) with 15 μl (1 mg ml^{-1}) of long-arm Photoprobe biotin (Vector laboratories).

2 Place the tube upright in crushed ice at a distance of 8–10 cm below a 275-W sun-lamp (General Electric). Irradiate for 30 min.

3 Add 15 μl of photobiotin to the reaction and irradiate for an additional 15 min.

4 Dilute the reaction (45 μl) to 150 μl with TE buffer, pH 8.0.

5 Perform five rounds of extraction with 150 μl of water-saturated 2-butanol.

6 Recover the aqueous phase (approximately 100–120 μl) and add 0.6 ml of target ss-cDNA eluted from the Qiagen Tip 5 column (see *Protocol 3*, step 9).

7 Co-precipitate target ss-cDNA and photobiotinylated driver poly(A)$^+$ RNA using equal volume of isopropanol (750 μl) at −20 °C overnight.[b]

[a] The ratio between ss-cDNA and biotinylated poly(A)$^+$ RNA should to be more than 1:20.

[b] During this step, tRNA must not be used.

4 Subtractive hybridization

In this chapter, we describe a method of subtractive hybridization in which biotinylated poly(A)$^+$ RNA has been used as a driver (see *Protocol 4*) and ss-cDNA as a target (see *Protocol 3*). What composition of hybridization cocktail is best? First, a hybridization cocktail has to contain an RNase inhibitor to prevent degradation of poly(A)$^+$ RNA during hybridization. Secondly, heteroduplex hybridization between target ss-cDNA and driver biotinylated poly(A)$^+$ RNA should proceed with high efficiency and speed.

Speed and efficiency of hybridization depend on the length of hybridizing nucleic acids, the nature (DNA or RNA; ss- or ds-) of nucleic acids, the composition of the hybridization cocktail, and the sequence of nucleic acids. Although the speed of hybridization between two nucleic acids can be calculated theoretically, the values achieved under experimental conditions are actually much slower. Moreover, when heterogeneous cDNA and RNA are manipulated, speed and efficiency of hybridization vary for each particular cDNA and cannot be predicted. Given these conditions, what hybridization cocktail is best?

Obviously, it would be best to avoid water-based reaction mixtures, although it is precisely this kind of hybridization cocktail which was used in the original work (1). Heteroduplex hybridization can be very effectively performed by vigorous shaking in phenol emulsion containing 1.5 M NaSCN (3). Furthermore, phenol is a well known inhibitor of DNase and RNase. Thus, a high level of subtraction was obtained when labelled cortex ss-cDNA (target) was hybridized with cerebellum ds-cDNA in the phenol emulsion. However, it was observed that biotinylated RNA does not hybridize well with ss-cDNA in the phenol emulsion. Therefore, we made a formamide-based hybridization cocktail. The best efficiency of heteroduplex hybridization was achieved when subtractive hybridization was performed for 40–48 h at 55–60 °C, and the final concentration of formamide in the hybridization cocktail was 20% (see *Protocol 5*).

Protocol 5

Heteroduplex hybridization and removal of ss-cDNA from ss-cDNA–poly(A)$^+$ RNA duplex

Equipment and reagents

- Hybridization cocktail (50 mM MOPS, pH 7.6, 0.2% SDS, 0.5 M NaCl, 5 mM EDTA, 20% formamide)
- Dilution buffer (50 mM MOPS, pH 7.6, 0.5 M NaCl, 5 mM EDTA)

Method

1 Co-precipitate the target ss-cDNA and biotinylated poly(A)$^+$ RNA (see *Protocol 4*, step 7) by spinning at 15 000 g for 1 h at 4 °C, wash the red pellet (the red colour of the pellet indicates that the RNA is successfully biotinylated) with 70% ethanol, dry, and

Protocol 5 continued

thoroughly dissolve in 5 μl of the hybridization cocktail. Add 50 μl of RNase- and DNase-free mineral oil (Sigma). Measure the levels of radioactivity of the mixture using a hand-operated monitor.

2. Heat the mixture for 2 min at 95 °C and incubate for 40–48 h at 58 °C.
3. Remove the tube from 58 °C water bath, and immediately add 100 μl of dilution buffer containing streptavidin (Boehringer Mannheim) at 3 U per 1 μg of biotinylated poly(A)$^+$ RNA.
4. Incubate for 10 min at room temperature.
5. Add 100 μl of phenol/chloroform (ratio 1:1), vigorously mix, spin for 3 min at room temperature, and remove carefully the aqueous phase.
6. Extract the organic phase with 25 μl of dilution buffer.
7. Add streptavidin at 1 U per 1 μg of biotinylated poly(A)$^+$ RNA to the aqueous phase (approximately 140–150 μl), and repeat steps 5 and 6 using 150 μl of phenol/chloroform. Measure the level of radioactivity of the final mixture using a hand-operated monitor to calculate the degree of subtraction.
8. Co-precipitate the rest of the target ss-cDNA using 5 μg[a] of freshly biotinylated poly(A)$^+$ RNA, 30 μl of 3 M sodium acetate, pH 5.0, and 2.5 volumes of ethanol. Repeat steps from 1 to 7.[b]
9. Precipitate 0.15–0.2 μl of subtracted target ss-cDNA using 10 μg of tRNA, 30 μl of 3 M sodium acetate, pH 5.0, and 2.5 volumes of ethanol. Keep the ss-cDNA under ethanol for 2 h at −20 °C.
10. Spin at 15 000 g for 30 min at 4 °C, wash with 70% ethanol, dry, and dissolve ss-cDNA pellet in 10 μl of water (*Figure 2*).

[a] We used three times less biotinylated driver poly(A)$^+$ RNA during the second round of subtractive hybridization.

[b] 60–100-fold final subtraction of target ss-cDNA can be achieved.

5 Construction of subtractive cDNA library

5.1 Converting subtracted ss-cDNA into ds-cDNA for the construction of a subtractive cDNA library

There are two fundamentally different approaches to convert ss-cDNA into ds-cDNA. Both approaches require the availability of primers, which are annealed to ss-cDNA and can be extended into the second chain of ds-cDNA with either Klenow fragment, or *E. coli* DNA polymerase I or Taq polymerase. The first approach is based on the capability of *E. coli* DNA polymerase I to extend RNA-primers annealed on ss-cDNA template into second cDNA chains. Thereafter ds-cDNA should be blunt-ended with T4 DNA polymerase. This approach requires ss-cDNA-RNA duplexes. RNase H activity creates RNA-primers and *E. coli* DNA polymerase extends these into the second strand of ds-cDNA (11). Obviously, this approach cannot be employed to convert subtracted ss-cDNA into ds-cDNA, because ss-cDNA-RNA duplexes do not exist after subtractive hybridization. Therefore, a second approach has to be used. 5′-ends of ss-cDNA have to have the same known sequence which is quite short (from 10 bp to 50 bp) to which primers can anneal. This 5′-end known sequence can be either synthesized with terminal deoxytransferase or ligated to ss-cDNA with T4 RNA-polymerase. T4 RNA-polymerase is an unreliable enzyme. Secondly, the separation of ss-cDNA from free non-ligated oligonucleotides is a very difficult process. Therefore, the use of terminal deoxytransferase to synthesize the same known 5′-end of ss-cDNAs is favoured. It has been observed that terminal deoxytransferase does not develop more than 30 bp oligo(dG) 5′-ends on ss-cDNA when dGTP is used at final concentration of 0.01–0.1 mM for 0.001–1 ng of ss-cDNA (see *Protocol* 6, steps 1 and 2).

5.2 Klenow fragment or Taq polymerase?

Typically, only a few nanograms (or even picograms) of target sequences remain after subtractive hybridization. Hence it is problematic to obtain a sufficient number of clones, because this requires large amounts of both driver and target mRNA. This requirement can be a major obstacle if we need to isolate mRNA from a few cells, such as parts of an embryo. Therefore, we preferred to employ Taq polymerase not only for conversion of subtracted ss-cDNA into ds-cDNA, but also for PCR-amplification of ds-cDNA as well. Non-specific T-primers and C-primers have been used to amplify subtracted G-tailed ss-cDNA (see *Protocol* 6).

5.3 Multistage PCR of ss-cDNA

The efficiency of the isolation of specific transcripts depends on the construction of a representative subtractive cDNA library from a small amount of target and driver tissues. The major problem with the use of PCR to amplify the whole pool of a subtracted cDNA is the over-representation of short transcripts, because short cDNAs are amplified more efficiently when they are a part of a cDNA pool. Thus, a 4 kb single cDNA clone is amplified three to five times less after 30 cycles of PCR than 0.5 kb cDNA. However, when 0.5 kb and 4 kb cDNAs enter into the composition of a heterogeneous cDNA pool, the 0.5 kb cDNA is amplified 20–40% more than the 4 kb cDNA after each cycle (12). One approach to overcoming this difficulty is to amplify larger transcripts in the cDNA pool for more PCR cycles (12).

It was observed that small cDNAs always contaminate large cDNAs on elution from an agarose gel. Thus, gel fractionation of large cDNA fragments can generate equally well 0.5 kb and 4 kb cDNAs when the starting amount of ss-cDNA pool is more than 5–10 ng (12). The approach described here was arrived at empirically, and uses three to four separate PCR amplification steps and allows us to amplify 0.5 kb and 4 kb cDNAs with equal efficiency from a starting amount of ss-cDNA pool as little as 5–10 pg. The first step of this approach is standard (see *Protocol* 6,

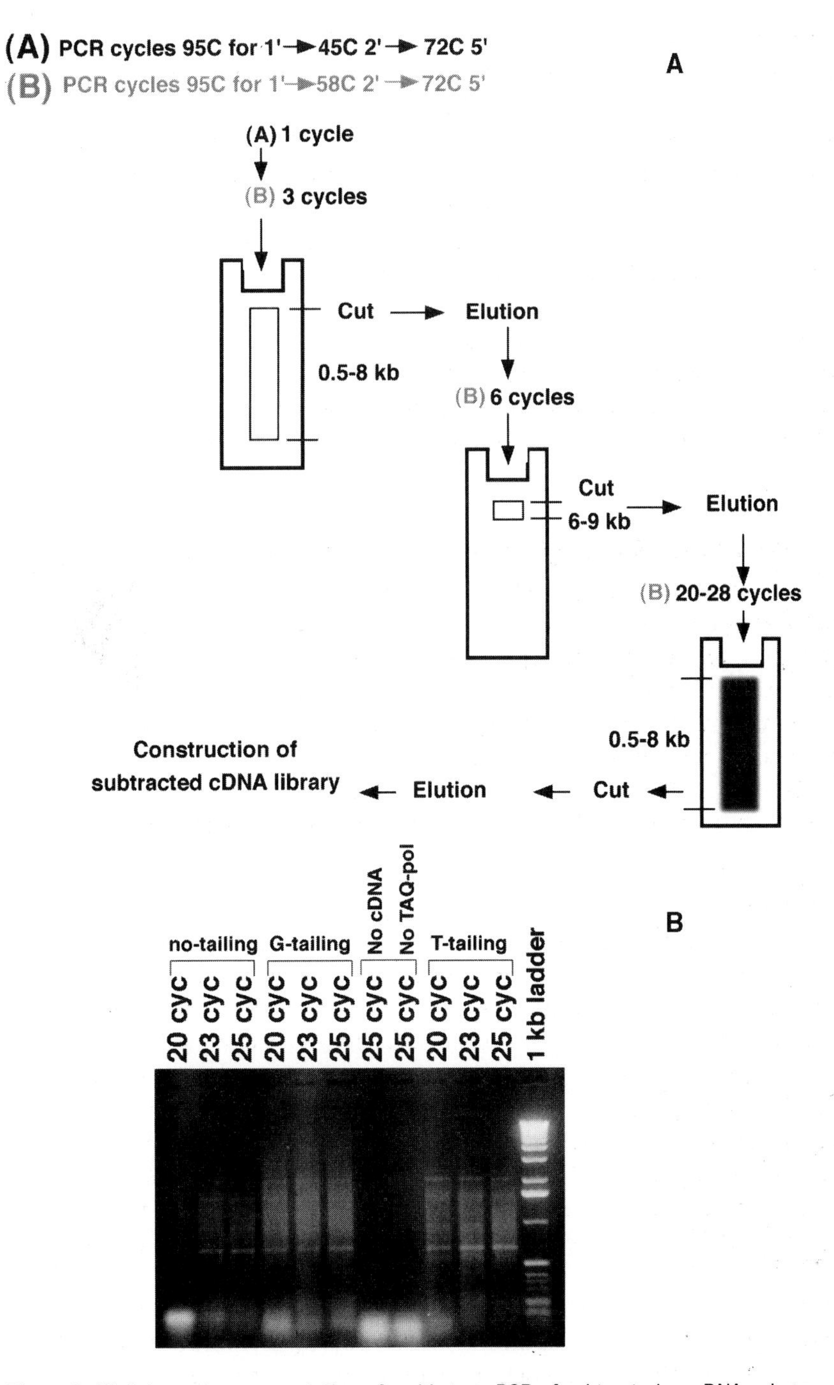

Figure 3 (A) Schematic representation of multi-stage PCR of subtracted ss-cDNA using T-primers and C-primers. B. Agarose (1%) gel electrophoresis of multi-step amplified subtracted ss-cDNA. Different deoxynucleotides were used for tailing of subtracted ss-cDNA. The numbers of cycles for the third step of multi-step PCR are shown.

steps 4–6) (*Figure 3A*). ss-cDNA is converted into ds-cDNA, then a 0.5–7 kb cDNA pool is separated from primer-dimers, T-primer-C-primer chimeras, and short cDNAs. The second step is also standard (see *Protocol* 6, steps 7–9). The pool of cDNAs is enriched with long cDNAs (5–7 kb) (*Figure 3A*). The protocol for the third step depends on the starting amount of subtracted ss-cDNA. If approximately 0.1–1 ng of the subtracted ss-cDNA is available for PCR amplification then 25–30 cycles is sufficient to generate a truly representative spectrum of transcripts (see *Protocol* 6, steps 10–12). Twenty to 25 cycles have been used when 1–10 ng of subtracted ss-cDNA is available (*Figure* 3). However, if only 1–100 pg of ss-cDNA is available, the third step has to be divided into two separate steps. First, the heterogeneous cDNA pool is amplified using 15–20 cycles, thereafter the pool of cDNA is again enriched by 5–7 kb cDNAs using gel fractionation. Finally, the enriched cDNA pool is amplified using 20–30 cycles. As precise amounts of subtracted ss-cDNA are very difficult to measure, the trial-and-error method should be used to achieve an optimal conditions for the third step.

Protocol 6

Amplification of subtracted target ss-cDNA with non-specific primers using PCR (*Figure 3*)

Equipment and reagents

- Standard 50 μl of PCR amplification mixture (26 μl of water, 5 μl of 10 × TAG buffer (Promega), 3 μl of 25 mM $MgCl_2$, 2 μl of dNTP solution containing 5 mM of each deoxynucleotide triphosphate, 2 μl of C-primer, 2 μl of T-primer, 0.5 μl (5 U μl^{-1}) of TAG-polymerase (Promega), and 10 μλ of cDNA)
- C-primer is 5′-AATTGCCA$(C)_{10}$-3′ (1 mg ml^{-1})
- 2% NuSieve gel in TAE, pH 7.5 (50 × TAE: 242 g Tris base (Sigma), 57.1 ml glacial acetic acid and 100 ml 0.5 M EDTA (pH 8.0))

Method

1. Add 4 μl of 5 × terminal deoxytransferase buffer (5 × TdT) (Promega), 1 μl of 1 mM dGTP, 5 μl of water to 10 μl of subtracted ss-cDNA (see *Protocol* 5, step 10). Mix and add 1 μl of terminal desoxytransferase (5–10 U μl^{-1}) (Promega). Gently mix and incubate for 1 h at 37 °C.
2. Extract the reaction with phenol/chloroform, pH 7.0–8.0, and precipitate with 3 M Na-acetate, pH 5.0, and 2.5 volumes of ethanol. Keep cDNA under ethanol for 2 h at −20 °C.
3. Spin at 15 000 g for 30 min at 4 °C, wash with 70% ethanol, dry and dissolve in 30 μl of water.[a]
4. Amplify the G-tailed cDNA in a standard PCR amplification mixture (see Equipment and reagents) under the following conditions: 95 °C, 1 min; 45 °C, 2 min; 72 °C, 4 min; 2 cycles; then 95 °C, 1 min; 58 °C, 1 min; 72 °C, 4 min; 2 cycles.

Protocol 6 continued

5 Run the amplified ds-cDNA on a 2% NuSieve gel.

6 Elute the ds-cDNA from 0.5 kb to 7 kb according to the standard procedure described in (12).

7 Repeat step 4 under following conditions: 95 °C, 1 min; 58 °C, 1 min; 72 °C, 4 min; 6 cycles.

8 Repeat step 5.

9 Elute the ds-cDNA from 5 kb to 7 kb according to standard procedure described in (12).

10 Repeat step 4 under following conditions: 95 °C, 1 min; 58 °C, 1 min; 72 °C, 4 min; 28 cycles.

11 Repeat step 5 and elute the ds-cDNA from 0.5 kb to 6 kb according to standard procedures described in (12).[b,c]

12 Precipitate ds-cDNA with 10 mg of tRNA, 20 mM Na-acetate, pH 5.0, and equal volume of 2-proponol. Keep ds-cDNA under ethanol for 1 h at −20 °C.

13 Spin at 10 000 g for 20 min at 15 °C, wash twice with 70% ethanol, dry and dissolve in 10 μl of water. The concentration of ds-cDNA is about 0.1–1 mg ml^{-1}.

[a] We have always used 1/3–1/5 part of cDNA solution for PCR-amplification.

[b] ds-cDNA can be visualised with ethidium bromide (0.5 mg ml^{-1}), while ds-cDNA is undetectable at step 5 and 8. Therefore, it must be eluted blindly.

[c] The ds-cDNA pool can be divided into two zones 0.5–2 kb and 2–6 kb, and separate subtractive cDNA library can be generated. It means the Protocol 7 has to be performed separately for both 0.5–2 kb and 2–6 kb ds-cDNA.

5.4 Choice of vector and generation of a subtractive cDNA library

An important aspect is the simplicity with which a cDNA library can be generated. ds-cDNA is blunt-ended and dephosphorylated after PCR. Therefore, phosphorylation followed by the ligation of ds-cDNA with adapters is necessary. However, there is a straightforward and simple method to produce 5′- and 3′-ends of ds-cDNA which are suitable for ligation of cDNA into vector. We have used adapter-primers containing *Not*I and *Eco*RI sites in T-primers and C-primers, respectively. The *Not*I sites were generated with *Not*I restriction enzyme (see *Protocol* 7, step 8), while *Eco*RI sites were produced with Klenow fragment employing its 5′–3′ exonuclease activity (see *Protocol* 7, step 5).

Are plasmid-based vector or λ-based vectors best to produce a subtractive cDNA library? A λ-based vector offers several advantages over a plasmid-based vector.

(1) Bacterial colonies bearing plasmids produce higher background than λ-phage plaques during highly sensitive differential screening that in turn seriously reduces the sensitivity of the whole method.

(2) The titre of DNA in λ-plaques is higher than in bacterial clones. Obviously, a high titre will increase sensitivity of screening.

(3) To screen a subtractive cDNA library, two to three replica filters derived from one dish are needed. It is critically important that DNA in phage plaques or bacterial clones be equally and effectively transferred from dishes on to replica filters. The methods of plaque-transfer on to replica filters are relatively simple, highly effective, and provide an equal distribution of DNA between at least three replica filters. On the other hand, the methods of bacterial colony transfer are more difficult to perform, and substantially different amounts of plasmid DNA are transferred on to first and second replica filters.

Two to three sequential screenings of a conventional cDNA library, which is plated at high density (10^5 clones/12 × 12 cm^2 dish), are usually required to identify an independent cDNA clone. Sensitivity of a screening is reduced in each cycle, if the same probes are successively used for all two to three screenings. To avoid a reduction in sensitivity, freshly synthesized probes are needed for each cycle of screening. However, in the case of differential screening, only one screening can be afforded because there is often a limitation of material for synthesizing probes. Therefore, a subtractive cDNA library has to be plated at low density (2–2.5 × 10^3 clones/12 × 12 cm^2 dish). Furthermore, plaques are larger when they are grown at low density, which in turn increases the sensitivity of differential screening.

Protocol 7

Construction of subtractive cDNA library

Equipment and reagents

- 1.5% agar dish (dissolve 25 g of Luria broth (Gibco BRL) in 1 litre of water. Add 15 g of L-agar (Gibco BRL) and autoclave at 120 °C for 30 min. Pour warm (50–60 °C) agar into several square 12 × 12 cm^2 and round 9 cm dishes
- 0.7% top agarose (dissolve 21 g of NZY (Gibco BRL) in 1 l of water. Add 7 g of agarose (Gibco BRL) and autoclave at 120 °C for 30 min

Method

1. Cut undigested λ UNI-ZAP DNA (Stratagene) with *Not*I and *Eco*RI. Mix 20 μl of λ UNI-ZAP DNA (10 mg), 8 μl of H-restriction buffer (Boehringer Mannheim), 46 μl of water, 3 μl of *Not*I (40 U μl^{-1}) (Boehringer Mannheim), and 3 μl of *Eco*RI (40 U μl^{-1}) (Boehringer Mannheim).
2. Incubate for 2 h at 37°C. Add 1 μl of calf intestinal phosphatase (CIP) (26 U μl^{-1}) (Boehringer Mannheim), and incubate for a further 40 min at 37 °C.
3. Add 120 μl of STE and extract the reaction mixture using 200 μl of Phenol, pH 7.5–8.0, then 200 μl of phenol/chloroform, pH 7.5–8.0. Precipitate DNA with 20 mM

Protocol 7 continued

Na-acetate, pH 5.0, and 2.5 volumes of ethanol. Keep λ-cDNA under ethanol for 1 h at −20 °C.

4. Spin at 10 000 g for 10 min at 4 °C, wash twice with 70% ethanol, dry and dissolve in 10 μl of water. Concentration of λ-DNA is 1 mg ml^{-1}.
5. Digest cDNA with Klenow fragment to create *Eco*RI sites. Add 5 μl of 10 × Klenow buffer (Boehringer Mannheim), 1 μl of 10 mM dGTP, 1 μl of 10 mM dCTP, 29 μl of water, 4 μl of Klenow fragment (5 U μl^{-1}) (Boehringer Mannheim) to 10 μl of amplified ds-cDNA (see Protocol 6, step 13). Incubate for 40 min at 37 °C.
6. Dilute the reaction with 150 μl of STE and extract the mixture using 200 μl of phenol/chloroform, pH 7.5–8.0. Precipitate ds-cDNA with 20 mM Na-acetate, pH 5.0, and 2.5 volumes of ethanol. Keep ds-cDNA under ethanol for 1 h at −20 °C.
7. Spin at 10 000 g for 20 min at 4 °C, wash twice with 70% ethanol, dry and dissolve in 10 μl of water.
8. Cut the ds-cDNA with *Not*I. Add 3 μl of H-restriction buffer, 15 μl of water to 10 μl of Klenow-digested ds-cDNA (step 7). Mix, add 2 μl of *Not*I (40 U μl^{-1}) (Boehringer Mannheim) and incubate for 2 h at 37 °C.
9. Dilute the reaction with 170 μl of STE and repeat steps 6 and 7.
10. Phosphorylate ds-cDNA before ligation into l-UNI-ZAP vector. Add 1.5 μl of 10 × T4 protein kinase buffer containing 1 mM ATP (Boehringer Mannheim), 2.5 μl of water to 10 μl of ds-cDNA with *Eco*RI and *Not*I ends. Mix, add 1 μl of T4 protein kinase (10 U μl^{-1}) (Boehringer Mannheim) and incubate for 1 h at 37 °C.
11. Dilute the reaction with 185 μl of STE and repeat steps 6 and 7, dissolve ds-cDNA pellet in 6 μl of water.
12. Mix 0.5 μl of 10 × ligation buffer (Boehringer Mannheim), 3 μl of ds-cDNA (see step 11) and 2 μl of λ-UNI-ZAP *Not*I–*Eco*RI vector (see step 4). Add 0.3 μl of T4-ligase (5 Weiss U μl^{-1}) (Boehringer Mannheim) and incubate for 16–24 h at 10 °C.
13. Package 1 μl of the ligation mix (see step 12) using Gigapack Gold (Stratagene) packaging kit.
14. Plate 1/1000 of SM solution (see Protocol 8) containing phage, which represents the packaged subtractive cDNA library, on 9 cm 1.5% agar dish using 0.7% top agarose and BB4 host cells.[a]
15. Finally, plate 2000–2500 plaques on 12 × 12 cm^2 square dishes.[b]

[a] This step is necessary to determine an exact titre of the subtractive cDNA library. Mix 1/1000 of phage solution in SM with 250 ml of BB4 cells in Luria Broth (D550 = 1). Incubate for 15 min at 37 °C, add 8 ml of 0.7% top agarose (40–45 °C), mix and pour on 9 cm 1.5% agar dish. Incubate dish for 8–10 h at 37 °C. Count phage plaques grown on bacterial lawn.

[b] We used four dishes to plate library. It is not reasonable to employ more than four dishes, because the sensitivity of differential and confirmatory screening (see next section) is likely to be reduced.

6 Differential and confirmatory screening of subtractive cDNA library

To detect clones bearing cDNA inserts, which constitute target-specific transcripts, special screening, named differential screening, should be performed. The principle behind differential screening is the separate hybridization of the cDNA library with a target-derived probe as well as with a driver-derived probe. The cDNA inserts, which are detected by hybridization with a target-derived probe, but not a driver-derived probe, could be considered as target-specific transcripts. The original method of differential screening of a conventional cDNA library has been described by Dworkin and Dawid (15).

6.1 Sensitivity of the screening

As has been mentioned, several factors are responsible for sensitivity of the whole experiment, but the screening of the subtractive cDNA library plays a crucial part in the sensitivity. The original technique of differential colony hybridization was only able to detect mRNA with an abundance of 0.05% or greater. This limitation can be major obstacle if one needs to isolate mRNAs with low abundance. What can be done to increase the sensitivity of the screening? How much [^{32}P]-radiolabelled dNTP and poly(A)$^+$ RNA for synthesis of probes are needed to detect clone representing mRNAs with an abundance as low as 0.0001–0.001%?

(1) The low sensitivity of the original differential screening is due to the low signal-to-noise ratio. To increase the signal-to-noise ratio, one of the three different strategies can be employed:
   (a) screening of a conventional cDNA library with subtracted probe,
   (b) differential screening of subtractive library,
   (c) screening subtractive cDNA library with subtracted probe (3, 5, 6).

(2) The amplification of plaques on the filters can give a 50–100-fold rise in sensitivity on differential screening. The method of plaque amplification on filters is performed essentially as described by Woo and colleagues (13).

(3) We observed that 1000 c.p.m. (Cerenkov) of a homogenous ss-cDNA per 1 ml of a filter-hybridization solution is sufficient to detect signal on a filter from amplified phages after 3 days expose with Kodak X-Omat film and intensifying screens at −70 °C. Thus, $100\,000 \times 1000 = 10^8$ c.p.m. ml^{-1} of heterogeneous cDNA synthesized from poly(A)$^+$ RNA should be used to reveal clones representing mRNAs of 0.001% abundance. Ten ml of a filter-hybridization solution is routinely used to cover four 12×12 cm^2 filters, and 500 μCi of pure [α-^{32}P] dNTP is equal to 10^9 c.p.m. This means 50% incorporation of 1 mCi isotope should be achieved during synthesis of cDNA probe from poly(A)$^+$ RNA to isolate rare (0.001%) transcripts.

(4) cDNA probe prepared from 0.8–1 μg of poly(A)$^+$ RNA and 1 mCi of [α-^{32}P] dATP using SuperScript reverse transcriptase results in a total incorporation of 10^8

c.p.m. Therefore, 8–10 μg of poly(A)$^+$ RNA isolated from both target and driver tissues is needed to incorporate 10^9 c.p.m., and by doing so rare transcripts (0.001%) could be isolated.

6.2 Artefacts and controls

Usually, many artefactual transcripts that are non-specifically expressed in the target tissue are isolated from a subtractive cDNA library by differential screening. Moreover, it takes a long time to complete a full analysis (see Section 7), and identify all artefactual clones. We suggested that additional confirmatory screening should be performed to avoid the isolation of artefactual clones (6). The subtracted amplified cDNA used for construction of a cDNA library (see *Protocol 6*) can be labelled with Klenow fragment and random-primers to screen the same subtractive cDNA library. Because the subtracted cDNA contains a relatively high proportion of target-specific transcripts, it will primarily identify cognate clones in the library. Obviously, not each clone identified by confirmatory screening is target-specific. However, if differential and confirmatory screening are performed simultaneously, then the isolation of artefactual clones will essentially be avoided before full analysis. Thus, 46 transcripts were analysed by Northern blotting; of these, 23 (50%) were apparently target-specific (6).

Protocol 8

Differential and confirmatory screening of subtractive cDNA library

Equipment and reagents

- Denaturing solution (0.5 N NaOH, 1 M NaCl)
- Neutralizing solution (0.5 M Tris, pH 7.0, 1.5 M NaCl)
- Fixing solution (0.4 N NaOH)
- 20 × SSC (175 g of NaCl and 104 g of Na-citrate (BDH) per 1 litre water)
- dNTP (Pharmacia) stock for labelling 10 μl of 100 mM dATP, 10 μl of 100 mM dGTP, 10 μl of 100 mM dTTP and 70 μl of water
- Denatured single-stranded salmon sperm DNA. Dissolve 1 g of salmon sperm DNA (Sigma) in water at 10 mg ml^{-1}, add 2.5 μl of 4 N NaOH. Boil for 20 min, neutralize reaction with 1 M Tris, pH 2.0
- 50 × Denhardt's solution. Dissolve 2 g of Ficoll (Sigma), 2 g of Polyvinylpyrrolidone (Sigma), 2 g of bovine albumin (Fraction V; Sigma) in 100 ml of water, and filter the solution throughout 0.2 mm filter (Sartorius)
- Filter hybridization buffer (4 × SSC, 5 × Denhardt's solution, 150 mg ml^{-1} of denatured single-stranded salmon sperm DNA (Sigma), 20 mg ml^{-1} of poly U (Pharmacia), 20 mg ml^{-1} of poly(C) (Pharmacia), 5 mM EDTA, pH 8.0, 0.5% SDS)
- SM (100 mM NaCl, 7 mM $MgCl_2$ (Sigma), 0.1% Gelatin (Sigma))

Method

1 Cut square (11.5 × 11.5 cm^2) Hybond N$^+$ (Amersham) filters. Pre-treat the filters with BB4 cells for amplification of phage plaques as described by Woo and colleagues

Protocol 8 continued

(13). Replicate subsequently each dish with three different treated filters (13). Make four to five holes with a sharp needle on the top agarose and filters; they will allow to align the three filters with each other and to identify the hybridized phage plaques.

2. Transport the replica filters with phage plaques on the top surface on fresh 1.5% agar dishes (12×12 cm^2). Grow 6–9 h at 37 °C.
3. Denature the phage DNA by placing filters containing amplified phage plaques on Whatman 3MM filters soaked with denaturing solution. Keep them for 10–15 min at room temperature.
4. Neutralize these filters by transferring them to Whatman 3MM filters soaked with neutralizing solution. Keep them for 15–20 min at room temperature.
5. Immobilize phage DNA to Hybond N$^+$ filters by keeping them for 20 min at room temperature on Whatman 3MM filters soaked with fixing solution.
6. Wash the filters three times with $5 \times$ SSC (100 ml per filter). Dry them. The filters are ready to hybridize with different probes.
7. Prepare three different probes for hybridization with filters (step 6).
   - Mix 50 μl of $5 \times$ RT buffer (Gibco BRL), 25 μl of 0.1 M DTT (Gibco BRL), 12.5 μl of dNTP stock for labelling, 3 μl of oligo(dT)12–18 (1 mg ml^{-1}) (Pharmacia), 100 μl of [α-^{32}P] dCTP (3000 Ci mmol^{-1}; 10 mCi ml^{-1}) (Amersham), 5 μl of 1 mM dCTP, 2 μl of RNasin (40 U μl^{-1}) (Promega), 10 μl of pre-heated (for 2 min at 70 °C, and on ice) target poly(A)$^+$ RNA (1 mg ml^{-1}), 42 μl of water and 2 μl of SuperScript reverse transcriptase (200 U μl^{-1}) (Gibco BRL). This is reaction A. Incubate reaction A for 90 min at 39 °C
   - Mix 50 μl of $5 \times$ RT buffer, 25 μl of 0.1 M DTT, 12.5 μl of dNTP stock for labelling, 3 μl of oligo(dT)12–18 (1 mg ml^{-1}), 100 μl of [α-^{32}P] dCTP (3000 Ci mmol^{-1}; 10 mCi ml^{-1}), 5 μl of 1 mM dCTP, 2μl of RNasin, 10 μl of pre-heated (for 2 min at 70 °C, and on ice) driver poly(A)$^+$ RNA (1 mg ml^{-1}), 42 μl of water and 2 μl of SuperScript reverse transcriptase. This is reaction B. Incubate reaction B for 90 min at 39 °C
   - Mix 25 μl of $10 \times$ Klenow buffer (Boehringer Mannheim), 12.5 μl of dNTP stock for labelling, 100 μl of [α-^{32}P] dCTP (3000 Ci mmol^{-1}; 10 mCi ml^{-1}), 20 μl of random-primers (1 mg ml^{-1}) (Pharmacia), 2 μl of denatured (for 5 min at 100 °C) amplified cDNA (1 mg ml^{-1}) (see Protocol 6, step 13), 90 μl of water and 2 μl of Klenow fragment (5 U ml^{-1}) (Boehringer Mannheim). This is reaction C. Incubate reaction C for 10 min at 25 °C and then 90 min at 37 °C.
8. Add 250 μl of STE, 4 μl (10 mg ml^{-1}) of tRNA and 35 μl of 3 M Na-acetate, pH 5.0 to reaction A, B, and C. Mix and precipitate with 550 μl of iso-propanol. Keep 30 min at −20 °C.
9. Spin at 15 000 g for 20 min at 15 °C, wash with 70% ethanol, dry and dissolve pellets in 500 μl of STE.
10. Add 35 μl of 3 M sodium acetate, pH 5.0, mix and precipitate with 550 μl of isopro-

Protocol 8 continued

panol. Keep 30 min at −20 °C. Repeat step 9. Dissolve labelled target ss-cDNA, driver ss-cDNA and amplified subtracted ds-cDNA in 300 μl of 0.2% of SDS. Measure activity of these probes; it should be approximately 109 c.p.m.

11 Prehybridize the filters (see step 6) for 4 h at 65 °C in the filter hybridization buffer (10 ml per filter). Repeat pre-hybridization with fresh filter hybridization buffer (10 ml per filter).

12 Hybridize each set of filters with denatured (for 5 min at 100 °C) target ss-cDNA, driver ss-cDNA and amplified subtracted ds-cDNA probes (see step 10) for 20–24 h at 65 °C in the filter hybridization buffer (2–2.5 ml per filter).

13 Wash the filters step by step with:

- 2 × SSC three times for 20 min at room temperature
- 2 × SSC, 0.5% SDS twice for 30 min at 68 °C
- 0.2 × SSC, 0.5% SDS twice for 1 h at 68 °C

14 Wrap the filters in Saran Wrap and expose the filters to Kodak X-Omat films for 3 days at −70 °C.[a]

15 Pick phage plaques that are hybridized with both target ss-cDNA and amplified subtracted ds-cDNA, but not with driver ss-cDNA (Figure 3). Store in SM solution for further analysis.

[a] Filters may be exposed for 1 day to determine signal-background ratio, and then either re-exposed for 3–4 days or re-wash (0.1 × SSC, 0.5% SDS for 1–2 h at 68 °C) and exposed for 3–4 days (see *Plate 1*).

Confirmatory screening has a further important advantage. Eight to 10 mg of target poly(A)$^+$ RNA is not always available for differential screening. To get around this obstruction, two replica filters obtained from a subtractive cDNA library can only be hybridized with a driver-derived cDNA probe and a probe which is used for confirmatory screening. The clones that are revealed by confirmatory screening but not hybridization with driver-derived probe can be picked for analysis.

7 Basic analysis of specific transcripts

To complete the process of isolation of specific transcripts, three different types of analyses should be executed.

(1) In order to eliminate redundant clones and select clones bearing the longest insert, the insert probes are random-prime labelled and cross-hybridized with each other. By this means the range of independent clones is isolated.

(2) The clones obtained from the subtractive cDNA library after differential screening may be expressed exclusively, preferentially or non-specifically in the target tissue. This question can be answered through Northern blot and/or

in situ hybridization analysis of the selected inserts. Northern blot analysis enables us to determine not only the degree of specificity of the selected clones, but also the sizes of transcripts. This in turn allows us to estimate how many of the selected clones are full-length. *In situ* hybridization offers a way to assess cell-specificity of the selected transcripts. Furthermore, *in situ* hybridization can be used when tissue is limiting for making a Northern blot.

(3) In order to determine the nature of transcripts, sequence analysis of the 5′-end of selected inserts should be carried out. Sequence analysis may assign each transcript to one of following groups: known tissue-specific transcripts, known transcripts expressed preferentially in target tissue, known non-specific transcripts, novel tissue-specific transcripts which contain known motif(s), novel transcripts with recognizable motif(s) expressed preferentially in target tissue and new transcripts with non-recognizable motif(s). For new transcripts with non-recognizable motif(s), the isolation of full-length cDNA is recommended, because sequencing of the full-length clones may provide fresh insights into the nature of these transcripts. Thus, sequencing of the full-length copy of an unknown clone enabled us eventually to classify it into the family of P2X receptors (6, 14).

8 Conclusions

It must be emphasized that the methods for construction and screening of subtractive cDNA library are still being developed, and it is worth carefully reflecting on what protocols will be useful and reliable for your task. The method described here uses ss-cDNA as target and biotinylated poly(A)$^+$ RNA as driver, PCR amplification of subtracted ss-cDNA, and both differential and confirmatory screening to achieve as much efficiency and sensitivity as possible. However, for less than 100 pg of target poly(A)$^+$ RNA, this method is not so reliable, because a multi-step PCR can not amplify a representative ds-cDNA from 0.1–1 pg of subtracted ss-cDNA. Secondly, for more than 10 mg of target poly(A)$^+$ RNA and 200 mg of driver poly(A)$^+$ RNA, this method is too cumbersome and is not optimal, because the amount of subtracted ss-cDNA (approximately 10–100 ng) is enough to construct representative cDNA library without using PCR. We suggest that the method developed by Belyavsky and colleagues (5) should be used for very small amounts of target poly(A)$^+$ RNA, and the method described by Duguid and co-workers (4) will be appropriate for moderate and large amounts of target poly(A)$^+$ RNA.

Possible modifications of these protocols include simplified PCR-amplification steps and the use of non-radioactive probes for screening of subtractive cDNA libraries. However, the choice of target and driver tissues, which is one of most important step to carry out successfully for this experiment, cannot be optimized, rationalized or calculated. Ideally, two or three separate experiments using the same target tissue and different driver tissue is the optimal approach.

Acknowledgements

We thank Drs Veronica Souslova, Kenji Okuse, and C-C. Chen for many helpful and fruitful discussions; Samantha Ravenall for critical review of the manuscript. We are grateful to Vladimir L. Buchman for his optimization of protocols to re-isolate γ-synuclein.

References

1. Hedrick, S. M., Cohen, D. I., Nielsen, E. A., and Davis, M. M. (1984). *Nature*, **308**, 149.
2. Chien, Y., Becker, D. M., Lindsten, T., Okamura, M., Cohen, D. I., and Davis, M. M. (1984). *Nature*, **312**, 31.
3. Travis, G. H. and Sutcliffe, J. G. (1988). *Proc. Natl Acad. Sci. USA*, **85**, 1696.
4. Duguid, J. R., Rohwer, R. G., and Seed, B. (1988). *Proc. Natl Acad. Sci. USA*, **85**, 5738.
5. Zaraisky, A. G., Lukyanov, S. A., Vasiliev, O. L., Smirnov, Y. V., Belyavsky, A. V., and Kazanskaya, O. V. (1992). *Dev. Biol.*, **152**, 373.
6. Akopian, A. N. and Wood, J. N. (1995). *J. Biol. Chem.*, **270**, 21264.
7. Chirgwin, J. M., Przybyla, A. E., MacDonald, R. J., and Rutter, W. J. (1979). *Biochemistry*, **18**, 5294.
8. Ninkina, N. N., Stevens, G. E., Wood, J. N., and Richardson, W. D. (1993). *Nucleic Acids Res.*, **21**, 3175.
9. Caterina, M. J., Schumacher, M. A., Tominaga, M., Rosen, T. A., Levine, J. D., and Julius, D. (1997). *Nature*, **389**, 816.
10. Sive, H. L. and St John, T. (1988). *Nucleic Acids Res.*, **16**, 10937.
11. Lapeyre, B. and Amalric, F. (1985). *Gene*, **37**, 215.
12. Belyavsky, A., Vinogradova, T., and Rajewsky, K. (1989). *Nucleic Acids Res.*, **17**, 2919.
13. Woo, S. L. C., Dugaiczyk, A., Tsai, M.-J., Lai, E. C., Catteral, J. F., and O'Malley, B. W. (1978). *Proc. Natl Acad. Sci. USA*, **75**, 3688.
14. Chen, C.-C., Akopian, A. N., Sivilotti, L., Colquhoun, D., Burnstock, G., and Wood, J. N. (1995). *Nature*, **377**, 428.
15. Dworkin, M. B. and Dawid, I. B. (1980). *Dev. Biol.*, **76**, 449.

Chapter 3
Differential display analysis of alteration in gene expression

Yong-Jing Cho and Peng Liang
Vanderbilt Cancer Center, Department of Cell Biology, Vanderbilt University School of Medicine, Nashville, TN, USA

1 Introduction

Temporal and spatial expression of the 100 000 different genes in the genome of a mammal is a highly regulated process which determines the normal development of an organism. The alterations in this process are therefore often the cause of developmental and pathological abnormalities such as cancer.

Two-dimensional protein gel electrophoresis was developed for the purpose of comparative studies. However, the limitation of this method became evident mostly due to its sensitivity problem, as at most only up to 2000 of 10–15 000 different types of protein expressed in any mammalian cell can be detected. Therefore, the majority of the genes, especially those expressed at lower levels, may not be represented. One other major pitfall of the method is the difficulty of being able to retrieve enough protein spots identified for future molecular characterizations, although there have been many recent technical advances in these areas (see Chapters 9 and 10 for further discussion of these issues).

More recently, numerous methodologies based on comparative studies at messenger RNA or cDNA levels have been developed. These include subtractive-hybridization (see Chapter 2), differential display (1, 2), representational difference analysis (RDA; see Chapter 4), serial analysis of gene expression (SAGE; see Chapter 7) and cDNA and oligonucleotide microarrays (see Chapter 6). This chapter focuses on the principle and practice of differential display, which has become one of the most widely used methods for cloning differentially expressed genes. In 1998 alone, nearly 400 publications appeared using differential display methods. In addition, the factors that affect the quality and reliability of differential display will be discussed, so the method can be used more successfully in the future.

2 Principle of differential display

To speed up the hunt for genes with regulated expression, differential display was developed with the aim to overcome limitations of previous methodologies.

To achieve this goal, the method has to be simple, sensitive, systematic, and reliable. The method has to be simple before it can be adopted easily and widely. The revolution in molecular biology, in fact, has been powered by mostly simple methodological breakthroughs such as recombinant DNA technology, DNA sequencing, and polymerase chain reaction (PCR). This simplicity also ensures the reliability of the method. The method has to be sensitive so it can be applied to biological systems where scarce biological samples are available. The method has to be systematic so a complete search of all the expressed genes is possible. With these in mind, differential display was developed with the combinatorial use of each of the three most powerful and simple molecular biological methods mentioned above, PCR, DNA sequencing gel electrophoresis, and cloning of the cDNA species of interest.

Since the original description of the differential display method in 1992, many modifications have been made to streamline and optimize the method (3). However, the essence of differential display methodology remains unchanged. The principle is depicted in *Figure 1*, incorporating the latest improvements using one-base anchored oligo(dT) primers (4). First, mRNAs are converted to cDNAs using three individual anchored oligo(dT) primers that differ from each other at the last 3′ non-T base. The use of anchored primers enables the homogeneous initiation of cDNA synthesis at the beginning of the poly(A) tail for any given mRNA. The resultant three subpopulations of cDNAs are further amplified and labelled with by incorporation of radiolabelled nucleotides during PCR in the presence of a set of arbitrary primers.

As a result, mRNA 3′ termini defined by any given pair of anchored-prime and arbitrary primer are amplified and displayed by denaturing polyacrylamide gel electrophoresis. Side-by-side comparisons of the resulting cDNA patterns between relevant RNA samples reveals differences which may represent mRNAs whose expressions have been altered. The following step-by-step protocol for differential display has been largely adopted from the instruction manual of the RNAimage™ kits from GenHunter Corporation.

3 Methods

3.1 Materials and equipment

(1) 5 × reverse transcription (RT) buffer: 125 mM Tris–Cl, pH 8.3, 188 mM KCl, 7.5 mM $MgCl_2$, and 25 mM dithiothreitol.

(2) Moloney murine leukaemia virus (MMLV) reverse transcriptase (100 u μl^{-1}).

(3) dNTP (250 μM).

(4) 5′-AAGCTTTTTTTTTTTG-3′ (2 μM).

(5) 5′-AAGCTTTTTTTTTTTA-3′ (2 μM).

(6) 5′-AAGCTTTTTTTTTTTC-3′ (2 μM).

(7) Arbitrary 13-mers (2 μM).

(8) 10 × PCR Buffer.

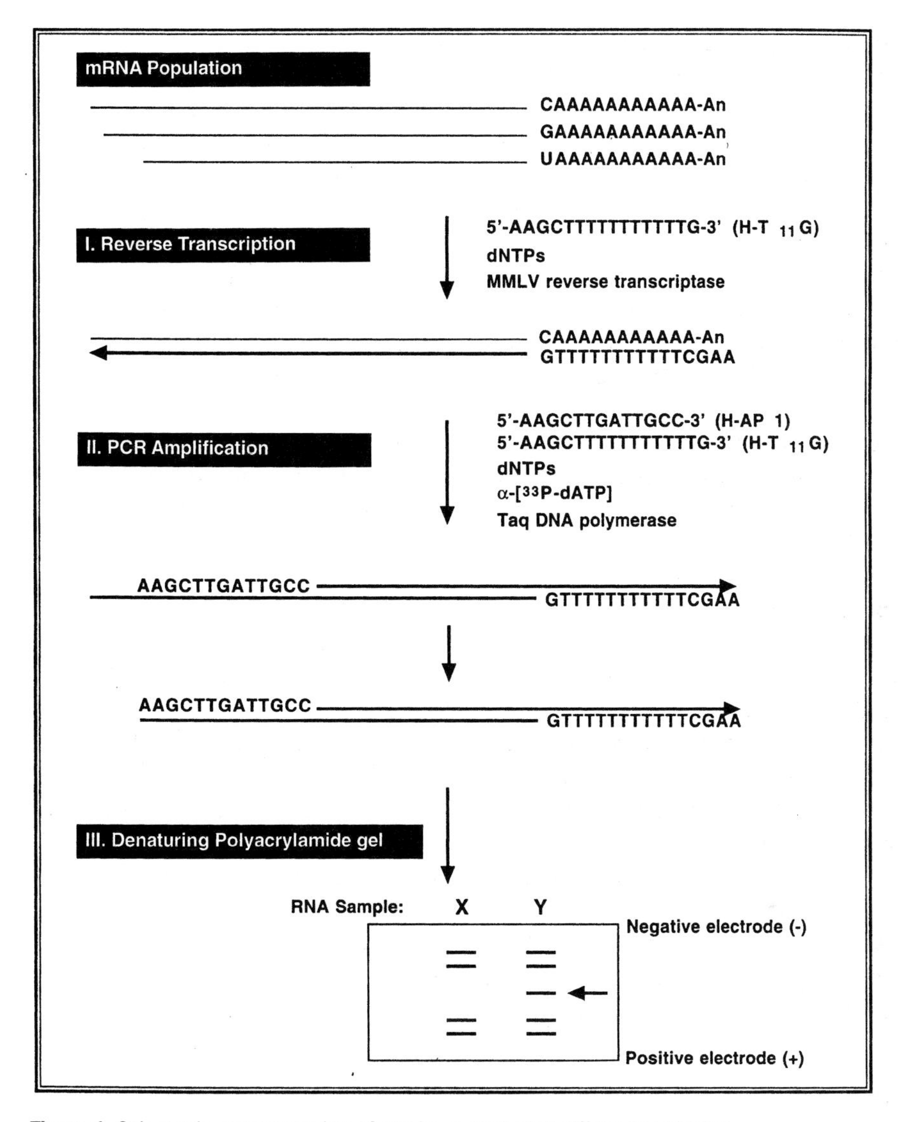

Figure 1 Schematic representation of one-base anchored differential display.

(9) dNTP (25 μM).

(10) Glycogen (10 mg ml^{-1}).

(11) dH_2O.

(12) Loading dye.

(13) Taq DNA polymerase (5 units μl^{-1}) (Qiagen).

(14) [α-^{33}P]dATP (>2000 Ci $mmol^{-1}$) (NEN-Dupont).

(15) RNase-free DNase I (10 u μl^{-1}).

(16) Thermocycler.

(17) Thin-walled PCR tubes.

(18) DNA sequencing apparatus.

Although most of individual components may be purchased separately from various suppliers, most of them can be obtained in kit forms from GenHunter Corporation.

3.2 Differential display—a detailed protocol

Purification of polyadenylated RNAs is neither necessary nor helpful for differential display. The major pitfalls of using the polyadenylated mRNAs are the frequent contamination of the oligo(dT) primers which give high background smearing in the display and the difficulty in assessing the integrity of the mRNA templates. Total cellular RNAs can be easily purified with one-step acid–phenol extraction method such as RNApure reagent (GenHunter Corporation) or RNAzol B (Biotecx). However, no matter what methods are used for the total RNA purification, trace amount chromosomal DNA contamination in the RNA sample could be amplified along with mRNAs thereby complicating the pattern of displayed bands. Therefore, removal of all contaminating chromosomal DNA from RNA samples is essential before carrying out differential display, as described in detail below, followed by a detailed protocol for carrying out differential display.

Protocol 1

DNAse treatment of total RNA to remove contaminating genomic DNA

Equipment and reagents

- DNase I (RNAse free)
- Phenol–chloroform
- Absolute and 70% ethanol
- 3 M sodium acetate
- DEPC-treated H_2O

Method

1. Incubate 10–100 μg of total cellular RNA with 10 units of DNase I (RNAse free) in 10 mM Tris–Cl, pH 8.3, 50 mM KCl, 1.5 mM $MgCl_2$ for 30 min at 37 °C.
2. Inactivate DNase I by adding an equal volume of phenol–chloroform (3:1) to the sample.
3. Mix by vortexing and leave the sample on ice for 10 min.
4. Centrifuge the sample for 5 min at 4 °C in an Eppendorf centrifuge.
5. Aspirate the supernatant, and ethanol precipitate the RNA by adding 1/10 volume of 3 M sodium acetate followed by 3 volumes of absolute ethanol, and incubate at −80 °C for 30 min.
6. Pellet the RNA by centrifuging at 4 °C for 10 min.

Protocol 1 continued

7 Rinse the RNA pellet with 0.5 ml of 70% ethanol (made with DEPC-H_2O) and air dry the pellet.

8 Redissolve the RNA in 20 μl of DEPC-treated H_2O.

9 Measure the RNA concentration at OD_{260} with a spectrophotometer by diluting 1 μl of the RNA sample in 1 ml of H_2O.

10 Check the integrity of the RNA samples before and after cleaning with DNase I by running 1–3 μg of each RNA on a 7% formaldehyde agarose gel.

11 Store the RNA sample at a concentration higher then 1 μg μl^{-1} at −80 °C before using for differential display.

Protocol 2

RT of RNA with anchored oligo(dT) primers

1 Set up three RT reactions for each RNA sample in three microfuge tubes (0.5 ml size), each containing one of the three different anchored oligo(dT) primers as follows. For 20 μl final volume: dH_2O (9.4 μl); 5 × RT buffer (4 μl); dNTP (250 μM) 1.6 μl; total RNA (DNA-free) 2 μl (0.1 μg μl^{-1}, freshly diluted); $AAGCT_{11}M$ (2 μM) 2 μl (M can be either G, A, or C); total 19 μl.

2 Program your thermocycler to 65 °C, 5 min −>37 °C, 60 min −>75 °C, 5 min −> 4°C.

3 Add 1 μl MMLV reverse transcriptase to each tube 10 min after at 37 °C and mix well quickly by finger tipping.

4 Continue incubation and at the end of the RT reaction, spin the tube briefly to collect condensation.

5 Set tubes on ice for PCR or store at −80 °C for later use.

Protocol 3

Differential display PCR

1 Set up PCR reactions at room temperature as follows. Twenty μl final volume for each primer set combination:

	μl
dH_2O	10
10 × PCR buffer	2
dNTP(25 μM)	1.6
Arbitrary 13-mer (2 μM)	2
$AAGCT_{11}M$ (2 μM)	2
RT-mix from *Protocol 2*	2 (it has to contain the same $AAGCT_{11}M$ used for PCR)

Protocol 3 continued

[α-^{33}P]-dATP (2000 Ci mmol^{-1})	0.2
Taq polymerase (5 u μl^{-1})	0.2
Total	20

Make core mixes as much as possible to avoid pipetting errors (e.g. aliquot RT-mix and AP-primer individually. Otherwise it would be difficult to pipette 0.2 μl of Taq.

2. Mix well by pipetting up and down. Add 25 μl mineral oil if needed.
3. PCR as 94 °C, 30 s –> 40 °C, 2 min –> 72 °C, 30 s for 40 cycles –> 72 °C, 5 min –> 4 °C. (For Perkin-Elmer's 9600 thermocycler it is recommend that the denaturation temperature be shortened to 15 s and the rest of parameters kept the same.)

Protocol 4

6% Denaturing polyacrylamide gel electrophoresis

1. Prepare a 6% denaturing polyacrylamide gel in TBE buffer according to standard techniques (this is the same type of gel as used for manual DNA sequencing). Let it polymerize at least for more than 2 h before using.
2. Prerun the gel for 30 min. It is crucial that the urea in the wells be completely flushed out right before loading your samples. For best resolution, flush every four to six wells each time during sample loading while trying not to disturb the samples that have been already loaded.
3. Mix 3.5 μl of each sample with 2 μl of loading dye and incubate at 80 °C for 2 min immediately before loading on to the 6% DNA sequencing gel.
4. Electrophorese for about 3.5 h at 60 W constant power (with voltage not to exceed 1700 V) until the xylene dye (the slower moving dye) reaches the bottom.
5. Turn off the power supply and blot the gel on to a piece of Whatman 3MM paper, as for orthodox sequencing gels.
6. Cover the gel with a plastic wrap and dry it at 80 °C for 1 h (do not fix the gel with methanol/acetic acid). Orient the autoradiogram and dried gel with radioactive ink or needle punches before exposing to an X-ray film. *Figure 2* shows a representative differential display of RNA from normal and ras oncogene-transformed rat embryo fibroblasts performed in duplicate to ensure reproducibility, which helps to eliminate false positives.

3.3 Recovery and amplification of cDNAs from differential display gels

After developing the film (overnight to 72-h exposure), orient the autoradiogram with the gel. Locate bands of interest either by marking with a clean pencil from underneath of the film or cutting through the film with a razor blade. Another way that has been found to work very well is to punch through the film with a

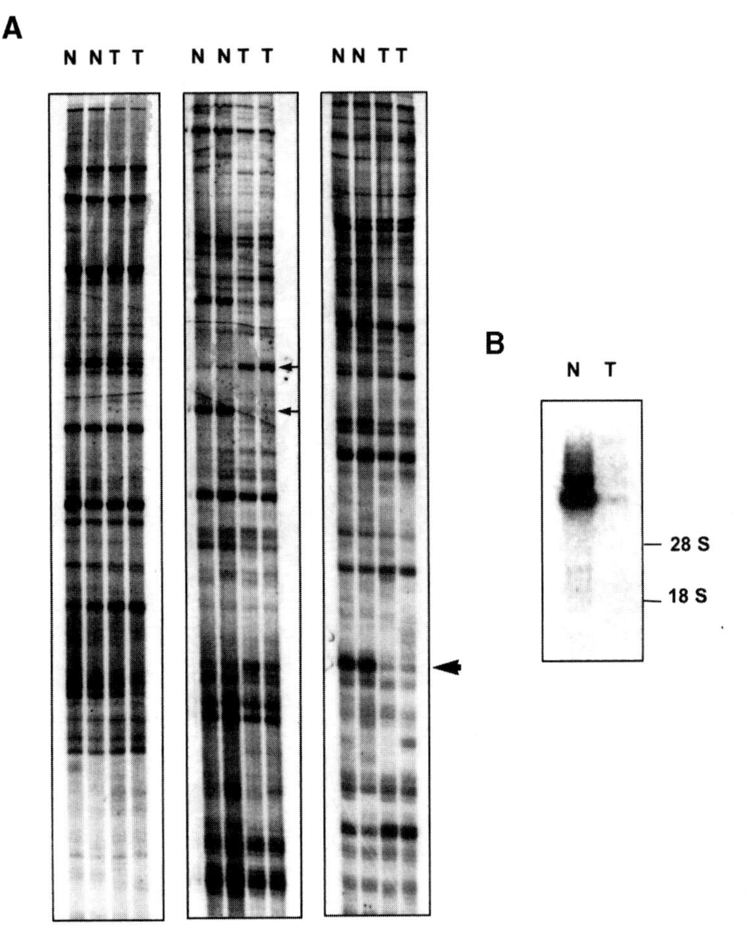

Figure 2 Differential display using one-base anchored oligo(dT) primer in combination with arbitrary 13-mers (4). RNA samples in duplicate from normal rat embryo fibroblast (N) and its derivative transformed oncogene H-ras (T) (9) were compared by differential display using a one-base anchored oligo(dT) primer in combinations with three arbitrary 13-mers (A). The arrows indicate consistent differences of which one marked by the large arrowhead was confirmed by Northern blot analysis (B).

needle at the four corners of each band of interest. (At all times, handle the dried gel with gloves and save it between two sheets of clean paper.) Once the gel is oriented on the autoradiogram, bands of interest should be recovered as described below.

Protocol 5

Recovery of cDNA from differential display gels

1. Cut out the located band with a clean razor blade.
2. Soak the gel slice along with the 3MM paper in 100 μl dH_2O for 10 min.

Protocol 5 continued

3 Boil the tube with tightly closed cap (e.g. with parafilm) for 15 min.
4 Spin for 2 min to collect condensation and pellet the gel and paper debris. Transfer the supernatant to a new microfuge tube.
5 Add in 10 μl of 3 M sodium acetate, 5 μl of glycogen (10 mg ml^{-1}) and 450 μl of 100% EtOH. Let sit for 30 min on dry ice or in a −80 °C freezer.
6 Spin for 10 min at 4 °C to pellet DNA.
7 Remove supernatant and rinse the pellet with 200 μl ice-cold 85% EtOH (you will lose your DNA if less concentrated EtOH is used!).
8 Spin briefly and remove the residual ethanol.
9 Dissolve the pellet in 10 μl of PCR H_2O and use 4 μl for reamplification. Save the rest at −20 °C in case of mishaps.

Protocol 6
cDNA reamplification

Reamplification should be done using the same primer set and PCR conditions except the dNTP concentrations are at 20 μM (use 250 μM dNTP stock) instead of 2–4 μM and no isotopes added. 40 μl reaction is recommended.

1 40 μl final volume for each primer set combination: dH_2O 20.4 μl; 10 × PCR buffer 4 μl; dNTP(250 μM) 3.2 μl; arbitrary 13-mer (2 μM) 4 μl; $AAGCT_{11}M$ (2 μM) 4 μl; cDNA template from RT 4 μl; Taq polymerase (5 units μl^{-1}) 0.4 μl; total 40 μl.
2 Amplify using the same PCR conditions as used for the original differential display PCR reaction.
3 Run 30 μl of the PCR sample on a 1.5% agarose gel stained with ethidium bromide.
4 More than 90% of reamplified cDNAs should be visible on the agarose gel.
5 Check to see if the size of your reamplified PCR products are consistent with their size on the denaturing polyacrylamide gel.
6 Extract the reamplified cDNA probe from the agarose gel using QIAEX kit (Qiagen) for Northern blot confirmation.
7 Save the remaining PCR samples at −20 °C for subcloning!
8 Clone the cDNA using the PCR-TRAP™ cloning system (GenHunter Corporation).
9 Verify that the cloned cDNA is differentially expressed by Northern blot (*Figure 2B*) or reverse Northern dot blot (5). The cloned cDNA can then be sequenced. Clone the full length cDNA by screening a cDNA library following the standard procedure.

4 Practical considerations for differential display

4.1 Factors affecting differential display

Although differential display has become one of the most popular methods used for cloning differentially expressed genes, its widespread use has also revealed

Table 1 Factors affecting the quality and reproducibility of differential display

Intrinsic factors
Primer designs
Anchored primers (one base vs. two bases)
Length of arbitrary primers (short vs. long)
Quality of reagents (enzymes, dNTP, and primers)
Extrinsic factors
Experimental design (controls)
Quality and quantity of RNA (integrity and contamination of DNA)
PCR tubes (thickness)
Isotopes [^{33}P]
Experimental set-up (core mixes, pipetting errors)
Thermocyclers (calibration)
Quality of the denaturing polyacrylamide gel electrophoresis
Criteria for picking bands
Experience of the researcher

some avoidable problem areas, which are known to affect the quality and reliability of the method. Some of the major factors are summarized below and dealt with in length in a recent review (6).

4.1.1 Causes of false positives

One of the most frequently asked questions by those who wish to use differential display is what the false positive rate is. There is really no clear answer to this very complex question, because it depends on so many factors (both intrinsic and extrinsic) of the differential display method. The intrinsic factors have been discussed above. The extrinsic factors such as the systems being compared, experimental designs, appropriate internal controls, criteria for picking bands, reaction set-up, type of PCR reaction tubes, type of thermocyclers, and, of course, the training and experience of a researcher, all can greatly contribute to the rate of false positives.

Taking one extrinsic factor, the system being compared, as an example, serves to illustrate some of the potential causes of false positives. If one is to compare gene expression between rat liver and rat brain, where it is known that nearly 50% of the genes expressed are brain specific, the number of false positives isolated is likely to be very low as the difference between samples compared are very large. However, if one is to compare gene expression between the same type of cells with and without a 30-min treatment of a drug or hormone in the hope of identifying immediate early genes; for example, the false positive rate will be much higher simply due to the fewer overall differences in gene expression.

Differential display itself can be perfectly reproducible if there are no intrinsic and extrinsic problems. For example, if you set up a 100 μl PCR reaction that is thoroughly mixed and aliquot 20 μl into each of five thin-walled PCR reaction tubes, the pattern of cDNA displayed among the five tubes can be perfectly the same. But if instead, five individual reactions are set up independently, the cDNA pattern amplified may not be the same simply due to pipetting errors. However, in practice, there is always going to be background 'noise' due to pipetting errors because you have to compare different samples. These few spurious bands or 'noise' can be as low as 0.1–1% of the cDNA bands displayed, which can be translated to 10–100 bands if you perform 240 PCR reactions, which statistically has more than 95% coverage of all the expressed genes in a cell. Therefore, given the systems compared, the false positive rate can be estimated simply as the signal (truly differentially expressed genes) to noise (spurious, non-reproducible bands) ratio. In summary, the smaller the total difference in gene expression, the higher will be the rate of false positives, but more relevant genes may be isolated.

The other often-ignored extrinsic factor is the appropriate internal controls. Assuming that you are looking for immediate early genes induced by a drug or hormone, you must make sure that the treatment truly worked. Such a decision can be based on either the phenotype of the cells, or more importantly, the induction of a known gene as a control. If this is not done, the differential display comparison is going to be flawed, because there may simply be no difference to begin with if the treatment did not work. In this case, if any differences are found, they are bound to be false positives.

4.2 Theoretical considerations

Unlike most of other methodologies, differential display is a sequence-dependent approach rather than prevalence-dependent approach to detect mRNA (7), which ensures its great sensitivity in detecting rare messages. However, one commonly asked question is how many primers have to be used to cover most of the expressed genes in a cells. *Table 2* provides a theoretical answer to this important question and in-depth explanation of the statistical treatment has been published elsewhere (8).

Table 2 Number of arbitrary 13-mers needed in combinations with all three one-base anchored primers to detect a given fraction of mRNA by DD

No. of arbitrary primers (n)	Reactions	Probability of detection $p = 1 - (0.96)n$
20	60	56%
30	90	71%
40	120	80%
80	240	96%

Acknowledgments

We thank GenHunter Corporation for the permission of adapting part of its protocol of RNAimage™ kit for differential display. This work was supported in part by grants CA76960 and CA74067 from the National Institutes of Health.

References

1. Liang, P. and Pardee, A. B. (1992). *Science*, **257**, 967–971.
2. Liang, P. and Pardee, A. B. (eds) (1997). *Methods in Molecular Biology*, Vol. 85. Humana Press, Totowa, New Jersey.
3. Liang, P. and Pardee, A. B. (1995). *Curr. Opin. Immunol.*, **7**, 274–280.
4. Liang, P., Zhu, W., Zhang, X., Guo, Z., O'Connell, R. P., Averboukh, L., Wang, F., and Pardee AB. (1994). *Nucleic Acids Res.*, **22**, 5763–5764.
5. Zhang, H., Zhang, R., and Liang, P. (1996). *Nucleic Acids Res.*, **24**, 2454–2456.
6. Liang, P. (ed.). (1998). *Methods—a companion to methods in enzymology*, Vol. 16. Academic Press, San Diego, California.
7. Wan, J. S., Sharp, S. J., Poirier, G. M.-C., Wagaman, P. C., Chambers, J., Pyati, J., Hom, Y.-L., Galindo, J. E., Huvar, A., Peterson, P. A., Jackson, M. R. and Erlander, M. G. (1996). *Nature Biotechnol.*, **14**, 1685–1691.
8. Liang, P. (1998). *Methods—a companion to methods in enzymology*, Vol. 16, pp. 361–364. Academic Press, San Diego, California.

Chapter 4
Representational difference analysis of cDNA

M. Hubank
Institute of Child Health, University College, London, UK

D. G. Schatz
Howard Hughes Medical Institute and Section of Immunobiology, Yale University School of Medicine, New Haven, CT, USA

1 Introduction

Accurate analysis of gene expression has become achievable in recent years thanks to the development of a variety of sophisticated techniques (1). In this chapter we describe the polymerase chain reaction (PCR)-based subtractive hybridization technique of cDNA representational difference analysis (cDNA RDA). cDNA RDA can be used to identify genes whose expression differs between two or more populations, and unlike gene array approaches, requires only basic laboratory equipment. cDNA RDA is a flexible, relatively inexpensive, and highly effective technique in which target cDNA fragments are sequentially enriched by favourable hybridization kinetics and subsequently amplified by PCR. The positive selection of differences and removal of common genes targets identification to only those genes which differ between populations, resulting in great sensitivity, and allowing application of the technique to very low amounts of starting material, including fluorescence-activated cell sorter (FACS) purified populations of cells.

cDNA RDA can be divided into three major phases: the generation of a PCR amplicon which is representative of the original mRNA from a given population; the two-step subtractive hybridization of different representative amplicons, during which amplified portions of differentially expressed genes are enriched and common sequences are depleted; and the cloning and screening of the resulting products. We present protocols for use with both high and low quantities of starting material, and detail a range of potential modifications to the method which permits tailoring to suit individual experiments. We also emphasize that prior consideration should be given both to the design and execution of experiments, and to the screening of differentially represented clones.

2 Principles of cDNA RDA

2.1 Genomic RDA and cDNA RDA

RDA was first described by Lisitsyn and his colleagues as a procedure for the identification of differences between two complex genomes (2). This was a significant advance in the field of subtractive cloning, for the first time successfully combining the advantages afforded by both subtractive hybridization and amplification. The procedure of genomic RDA, and the hybridization kinetics involved in the technique have been amply described elsewhere (2, 3), and this chapter will focus on the applications of RDA to cDNA.

cDNA RDA is a modification of RDA in which the aim is to identify the dynamic differences between expressed populations of genes, rather than the static differences between whole genomes (4). Genomic RDA requires a reduction in the complexity of starting material by the generation of a subpopulation of amplified restriction fragments that represent approximately 10% of the genome. This is because the sequence complexity of genomic DNA is too high to permit complete hybridization during the subtraction stages of the procedure. In cDNA RDA, the starting material is derived from messenger RNA (mRNA) rather than genomic DNA, and targets the subtraction to just those genes that are expressed at the time of isolation. Because the complexity of cDNA is substantially less than genomic DNA, simplification of the material is not strictly required. However, in order to take advantage of the subsequent enrichment of differences afforded by PCR, it is still necessary to generate an amplified population of cDNA restriction fragments, which is known as a 'representation'.

2.2 Representations

Ideally, a representation should contain at least one amplified product from every expressed gene. To achieve the highest degree of representation of cDNA while maintaining the amplifiability of as many fragments as possible, double-stranded cDNA is derived from the original mRNA by reverse transcription (RT) and second strand synthesis, and digested with a four-cutter restriction enzyme (typically *Dpn*II), which maximizes the generation of amplifiable fragments. After digestion, oligonucleotide adaptors (R-adaptors) are ligated on to the cut cDNA and PCR is performed. The resulting product is limited in its complexity by the ability of each product to amplify within the mixture. Templates which are either too large (>1.2 kb) or too small (<130 bp) do not amplify efficiently under what are effectively competitive conditions, and the product therefore only 'represents' the amplifiable proportion of the digest. The amplified representation then serves as the starting material for successive rounds of subtraction and amplification.

Statistical investigation has shown that *Dpn*II cuts cDNA to produce fragments with an average length of 286 bp, so it is likely that the vast majority of cDNA species will contain at least one potentially amplifiable fragment. Mathematical models and database analysis predict that a *Dpn*II representation typically includes at least one fragment from over 86% of expressed mammalian genes. This proportion can, in principle, be elevated to 94% if the primer used to generate first strand cDNA is designed to contain a *Dpn*II site. On this basis, the chances of missing one differentially expressed gene in a cDNA RDA subtraction would be about 10%, and those of missing three or more genes only one in a thousand. The majority of genes that are not represented are encoded by transcripts whose total lengths, including 3′ and 5′ untranslated regions, are below 500 bp.

These figures do not, however, take into account the efficiency of amplification of individual molecules within the mixture. Some species are likely to amplify better than others, while a very small proportion of fragments will not amplify at all, possibly because of inappropriate sequence close to the primer binding site. Comprehensive data on the composition of representations is not yet available, but it is our experience that almost all previously known fragments that fall into the amplifiable range do in fact amplify, although on occasions it is necessary to optimize conditions. The majority of genes contain more than one *Dpn*II restriction fragment, and so the likelihood is if one fragment of a gene fails to amplify, another one will, and the gene will be identified.

It is crucial that amplification of representations is kept within the linear range if the relative proportions of species are to be maintained with respect to the starting mRNA population. As long as pilot reactions are performed to determine the most appropriate number of cycles, representations seem to reflect mRNA composition remarkably well, and can even be used to probe gene microarrays (5, 6). Use of representations in place of RNA in microarray screening is a particularly exciting development because in many situations RNA for probing gene arrays is likely to be in short supply.

Maintenance of proportionality within representations is also critical if cDNA RDA is to be used to detect relative, rather than absolute differences. It is important to remember that while cDNA RDA is highly reliable for the discovery of target genes whose abundance differs markedly between the representations (i.e. at least 10-fold), the detection of relative differences of a lower magnitude, although possible, is not so predictable. Where fragments are present in both driver and tester at different levels, their amplifiability after subtraction depends heavily on the kinetics of hybridization of individual molecules within the complex mixture. It is not yet possible to forecast either the proportion of up-regulated fragments that will be enriched, or whether a particular sequence will be detectable at a given stringency of subtraction.

2.3 PCR-coupled subtractive hybridization

2.3.1 First difference product (DP1)

cDNA RDA differs from display-based techniques in that fragments which are common to both populations should be eliminated from the experiment while the abundance of differentially expressed genes is increased exponentially. This approach simplifies the subsequent identification and analysis of products and greatly reduces the proportion of false positives. cDNA RDA brings about positive selection of expressed differences by a process involving the subtraction of representations by hybridization, during which common material is selectively degraded, and an amplified 'difference product' is produced.

The representation derived from the material containing the target genes is designated the tester, while the control material is known as the driver, which will be used to 'drive out' fragments common to both samples. A tester is generated by removal of the R-adaptor used to generate the representation, and the ligation of a new adaptor consisting of unphosphorylated 12 and 24 base oligonucleotides. In this ligation, as in all cDNA RDA adaptor ligations, the 12-mer only provides the appropriate end structure to permit ligation of the 24-mer to the digested DNA, and is not itself covalently linked to the DNA. The driver and tester are then mixed at a driver/tester ratio of 100:1, melted and allowed to anneal under optimum conditions.

Three types of hybrids can be formed. The abundant driver/driver hybrids form most frequently, but these lack adaptors, cannot generate a primer binding site during the initial fill-in reaction, and are therefore incapable of amplification in the subsequent PCR. Driver/tester hybrids are the next most common product. Assuming that a fragment is present at equal levels in the two representations, then each strand of tester is 100 times more likely to anneal to a complementary strand from the driver than to a complementary strand from the tester. Because the driver strand is unable to generate a primer-binding site, these fragments can only amplify in a linear fashion. If, on the other hand, a target fragment in the tester has no complementary strand in the driver because it is not expressed in that population, then it can only anneal to a complementary strand from the tester itself, generating fragments with 24-mer oligos ligated on to each 5′ end. Residual 12-mers are then melted away, and a fill-in reaction is performed to generate the primer binding site (complementary to the 24-mer) for the following PCR. The primer is the 24-mer itself, and because only tester/tester hybrids have the primer binding site on each 5′ end, they alone will amplify exponentially. Part way through the PCR amplification, single-stranded DNA is degraded with mung bean nuclease, thereby eliminating both driver (which denatures during the PCR) and unamplified tester DNA. The resulting first difference product (DP1) is substantially depleted for fragments that are common to the starting representations, and modestly enriched for differences. It is necessary to maintain the representation of rare transcripts while avoiding biasing the population of amplifiable fragments in favour of more abundant species, so the hybridization ratios of driver and tester are delicately balanced, and the numbers of PCR cycles is kept within the linear range.

2.3.2 Second difference product (DP2)

Differences may already be apparent in the first product of subtraction and amplification (DP1), but because of random annealing events, many amplified tester/tester hybrids will not represent true differences at this stage. For this reason a second round of subtraction is performed, in a similar manner to the first, generating a second difference product (DP2). However, because of the enrichment in target sequences achieved at DP1, the ratio of driver to tester in the second subtraction is usually increased from 100:1 to 800:1, increasing the selection against gene fragments present in both populations. Owing to the low complexity of the material compared with genomic DNA, a third round of subtraction is seldom required in cDNA RDA, and in some cases can actually result in the loss of target sequences. After two rounds of subtraction and amplification, the difference products are visible on agarose gels as clear bands and can be easily subcloned into a suitable vector for primary screening.

3 Experimental design

cDNA RDA has the capability to pinpoint specific differences while excluding unwanted information. Careful experimental design is essential if full advantage is to be taken of these merits. In planning cDNA RDA experiments consideration of secondary screening methods should be uppermost in the minds of researchers. Secondary screening involves the functional investigation of genes once they have been established as true and reproducible differences by primary screening. True differences can often be surprising and more abundant than expected, and a convenient model for assessing their physiological relevance is crucial. In most cases this is likely to involve an *in vitro* system, at least as an intermediary, before whole organism work is undertaken.

Once test systems for the difference products have been established, the subtractions themselves should be designed to avoid the identification of differences which though real, are not of interest to the researcher. cDNA RDA is extremely sensitive and even very low levels of expression will be detected. Sometimes this is an advantage, but on other occasions it can result in the isolation of differences unconnected with the phenotype under investigation. The sensitivity of cDNA RDA can be particularly problematic when cell lines are the subjects of subtraction. It is common to find differences in gene expression between subclones, and even freeze-downs of lines. Such differences often appear arbitrary, and can confuse and confound research unless they can be rapidly identified and eliminated from the investigation. Subtractions should preferably be performed on cell lines separated in their origin by the shortest time and the fewest possible manipulations. Where, for example, the transcriptional effects of a complementing gene transfected into a mutant line are to be analysed by cDNA RDA, the resulting difference products should be screened against several separate complemented lines to rule out random events. In addition, the state of growth of cells should be identical to avoid the cloning of stress or cell cycle related genes where these are not the intended targets. If *in vivo* systems are the targets of investigation, subjects should be as closely related as possible, and great care should be taken with the dissection of tissues from which representations are to be prepared. A tiny amount of contaminating tissue can result in a lot of gene differences!

Subtractions can also be very effectively targeted using a number of modifications to the cDNA RDA procedure. Competitor fragments can be included in the driver to prevent the amplification of known differences, various combinations of drivers or testers can be employed, and the driver/tester ratios can be manipulated to permit the detection of relative, as well as absolute differences. In most circumstances it should be possible to use a combination of these approaches, which are described in more detail in Section 7, to reduce irrelevant differences to a minimum.

4 Application of cDNA RDA

4.1 General considerations

4.1.1 Time and cost

The length of time a subtraction takes, and the amount of material used will vary with the experiment and the researcher. As a guide, a basic cDNA RDA experiment involving a reciprocal subtraction of two representations will take a proficient operator approximately 2 weeks from isolation of RNA to cloning the DP2. A particularly committed worker could reduce this to 8 days, while most first time users are likely to need 3 weeks or longer. More complex experiments involving the use of mixed drivers, testers, or competitors will require proportionally more time. In practise, the length of time taken to perform the subtraction is small when compared with the time devoted to screening the products. This is why it is so important to target the subtractions to the identification of only the genes that have the highest likelihood of being related to the phenotype in question.

Basic cDNA RDA is relatively inexpensive when compared with array based approaches. No special equipment is necessary beyond a refrigerated microcentrifuge and a PCR machine. A typical reciprocal subtraction to DP2 requires approximately 50 μl of *Dpn*II, 30 μl of *Taq* DNA polymerase, six oligonucleotides (80 μg in total), part of one poly(A)$^+$ RNA isolation kit, and one DNA purification kit, together with small quantities of other common molecular biology reagents and laboratory consumables. An experiment requiring only one tester will consume only half this amount, while more representations or the addition of a competitor raise the cost proportionally.

4.1.2 Precautions

cDNA RDA is extremely sensitive, so all possible precautions should be taken to avoid the cross-contamination of materials. Cross-contamination of templates during cDNA RDA can cause the detection of false positives or the failure to detect true differences. 'PCR-clean' methodology should be observed throughout the protocol, including the use of barrier tips to reduce aerosols when handling any potential templates and the aliquoting of enzymes, buffers, and oligonucleotides. We recommend that all PCR reactions be set up in an airflow cabinet. Ensure that all reagents for RNA isolation and processing are RNase-free, and that only the highest quality reagents are used throughout the procedure.

4.1.3 Source material

The source material for cDNA RDA can be anything from which it is possible to reliably prepare RNA. cDNA RDA has been used to identify the targets of transcription factors in animal (7, 8) and plant cells (9); the transcriptional targets of hormones in animal tissue (10); the changes in transcription associated with development in cells purified from the immune system (11, 12); infection of tissues by viruses (13) and changes in gene expression associated with cancer (14, 15). cDNA RDA has also been applied to prokaryotic RNA, where the procedure has been considerably modified to compensate for the lack of polyadenylation of most bacterial transcripts (16). FACS purified cell populations make good starting material for cDNA RDA providing that clearly differentiating markers are used in the sorting, and that gates are set far apart. Populations should be checked by reanalysis, and cross-contamination should be less than 1%.

In principle it is possible to create amplicons from minuscule quantities of material, including single cells. However, problems can arise in maintaining a true representation of starting mRNA at very low cell numbers, and great care should be taken to confirm the validity of any products generated from very small amounts of cDNA. The protocols presented here are suitable for use with 10^4 cells and above.

4.2 Isolation of RNA

4.2.1 Total RNA

In common with all difference cloning techniques, successful cDNA RDA is heavily reliant on the quality of the starting material. It is therefore important to treat the source material identically, whichever method is chosen for isolation of RNA. If representations are to be derived from cells in culture, it is critical that the two cell populations are grown under identical conditions and are harvested at similar densities. RNA should be isolated from similar numbers of cells, using the same procedures, and preferably at the same time.

Total RNA can be isolated from large or small amounts of starting material. The advantage of using larger quantities of material is that sufficient remains to prepare Northern blots for screening, and that easily measurable quantities of both RNA and cDNA are produced which can be used to assess both the quality of the products and the progress of the experiment. Many reliable methodologies exist for the preparation of total RNA from large quantities of tissues and are available commercially or in standard laboratory manuals.

Frequently, however, only very small quantities of samples are available. In these circumstances, *Protocol 1* should be used. This method is optimized for use with between 10^4 and 10^7 cells and is particularly useful for FACS sorted populations of cells.

Once the RNA isolation process has begun, it should be continued until the total RNA is resuspended in water. Leaving the RNA for prolonged periods in Trizol or in ethanol containing salt is detrimental to the isolation procedure and to subsequent reactions. Total RNA can be stored in ethanol, but never in ethanol with salt (e.g. sodium acetate).

Protocol 1

Preparation of total RNA from <10^7 animal cells[a]

Equipment and reagents

- Refrigerated microcentrifuge
- Polytron homogenizer (for *Protocol 1C* only)
- Trizol reagent (Gibco BRL (Life Technologies Ltd))
- Chloroform; isopropanol; 75% ethanol and glycogen (Boehringer Mannheim)
- RNAse-free water

A From cell suspensions

1. If the cells are grown *in vitro* in suspension, grow them to a suitable density and take an aliquot for counting. If FACS sorted cell populations are to be used, sort up to 10^7 into ice-cold medium in FACS tubes.
2. Pellet up to 10^7 cells in a suitable tube by centrifugation at 400 g for 5 min at 4 °C. Pour off the supernatant and disrupt the pellet by flicking the base of the tube.
3. Resuspend the cells in Trizol reagent (Life Technologies). Use 0.5 ml of Trizol reagent for any number of cells up to 10^7 if the cells are small (e.g. lymphocytes). For larger cells (e.g. fibroblasts) use 1 ml for between 5×10^6 and 10^7 cells and double the following volumes accordingly.
4. Transfer the cells to a 1.5 ml microcentrifuge tube and incubate in Trizol for 5 min at room temperature, vortexing occasionally to ensure that all cells are completely disrupted.
5. To each 0.5 ml of cells in Trizol, add 100 μl of chloroform, mix well by shaking for 15 s and leave at room temperature for 2–3 min, then centrifuge at 14 500 g for 15 min at 4 °C.
6. Take the aqueous upper phase (which contains the RNA) to a new tube, add 250 μl of isopropanol,[b] and keep it at room temperature for 10 min. It is possible to include 1 μl of 10 mg ml^{-1} glycogen as a carrier in the precipitation. Spin at 14 500 g for 20 min at 4 °C.
7. Pour off the supernatant, re-spin for 1 min and remove the residual supernatant with a drawn-out Pasteur pipette. Wash the pellet with 1 ml 75% ethanol, vortex, and centrifuge at 9000 g for 5 min at 4 °C. Remove the ethanol with a drawn-out Pasteur pipette, allow the pellet to air dry for 1 min, and resuspend it in 75 μl of double-deionized, RNase-free water.

B From adherent cell monolayers

1. Grow adherent cultures to a density that is known to correspond to approximately 10^7 cells per flask or dish.

Protocol 1 continued

2 Aspirate the medium and, without washing, add 1 ml of Trizol reagent per 25 cm^2 flask. Swirl the flask for 5 min, or until lysis is complete[c].

3 Remove the Trizol-lysed cells to a 1.5 ml microcentrifuge tube and resume *Protocol 1A* at step 5.

C From small quantities of tissues

1 Dissect the required tissues as rapidly as possible. Remove all extraneous material[d]

2 Place the tissue directly into 1 ml of Trizol in a 15 ml Falcon tube.

3 Immediately homogenize the tissue on ice using a Polytron homogenizer set at a level which disrupts the tissue completely within 30 s, but avoids causing foaming.

4 Transfer the lysed tissue in 1 ml of Trizol to a microcentrifuge tube and resume *Protocol 1A* at step 5.

[a] For greater than 10^7 cells or the tissue equivalent, follow the protocol supplied by the manufacturer of Trizol Reagent., or use the equivalent method of Chomczynski and Sacchi (17).

[b] Do not leave isopropanol precipitations on ice for more than 1 h at any point in the protocol, and never place isopropanol precipitations in a freezer as this produces a heavy salt precipitate, which will inhibit later reactions.

[c] Alternatively, trypsinize the cells, wash them with phosphate-buffered saline, perform a count, and continue as in *Protocol 1A*, step 2.

[d] The quantity of material can either be estimated either by weighing tissue, by assaying total protein from the Trizol phenol phase, or by measuring RNA yield on a Northern blot or by RT-PCR.

4.2.2 Poly(A)$^+$ RNA (mRNA)

Even the most careful total RNA preparation contains contaminating genomic DNA that may compromise the cDNA RDA procedure. Genomic DNA should be removed prior to isolation of poly(A)$^+$ RNA with DNase I. This step is particularly important if very small amounts of starting material are used, as more PCR cycles will be necessary in creating a representation. If total RNA, rather than poly(A)$^+$ mRNA, will be the substrate for the generation of cDNA, successful DNase I treatment becomes critical. Where large amounts of starting material are employed, however, and poly(A)$^+$ RNA can be double selected with oligo(dT) cellulose, DNase treatment is not essential.

In principle, total RNA can be used as a template for cDNA RDA; however, mRNA generates more reproducible results and should, wherever possible, be isolated prior to making representations. There are several commercially available kits capable of preparing adequate mRNA from small quantities of material. We have found that a modified version of the Oligotex mRNA separation protocol (Qiagen) generates RNA from which reliable representations can be obtained using as few as 10^4 EL-4 mouse thymic lymphoma cells or 10^5 resting lymphocytes, which are smaller and yield less RNA. For large quantities of material it is more economical to column purify poly(A)$^+$ RNA using oligo(dT) cellulose which can subsequently be regenerated.

Protocol 2

Removal of residual genomic DNA[a]

Equipment and reagents

- Refrigerated microcentrifuge
- 0.1 M dithiothreitol (DTT); 3 M sodium acetate pH 5.3; 50 mM $MgCl_2$
- RNase Inhibitor (Promega; 20–40 U μl^{-1})
- DNase I (RNase-free) (Boehringer Mannheim; 10–50 U μl^{-1})
- Phenol/chloroform/isoamyl alcohol (25:24:1) (P/C/I);chloroform/isoamyl alcohol (24:1) (C/I); isopropanol; 70% ethanol
- RNase-free water

Method

1 Mix in order: 10 μl of 0.1 M DTT, 10 μl of 50 mM $MgCl_2$, 3.3 μl of 3 M Sodium acetate, pH 5.3, 0.5 μl of RNase inhibitor and 1 μl DNase I (RNase free). Add the mix to 75 μl of total RNA and incubate at 37 °C for 15 min.

2 Extract twice with 100 μl of P/C/I.[b] Extract once with C/I and move aqueous phase to a new tube.

3 Use directly for poly(A)$^+$ RNA isolation. Alternatively, precipitate by adding 7 μl of 3 M sodium acetate and 250 μl of 100% ethanol. Mix and incubate at −20 °C for 20 min. Centrifuge at 14 500 g for 15 min at 4 °C, wash with 70% ethanol and resuspend the pellet in 20 μl of RNase-free water.

[a] This step is particularly important if total RNA, rather than poly(A)$^+$ mRNA, will be the substrate for the generation of cDNA.

[b] The phenol extraction is performed twice to remove all traces of DNase I.

Protocol 3

Preparation of poly(A)$^+$ RNA (mRNA) from $<10^7$ cells[a,b]

Equipment and reagents

- Refrigerated microcentrifuge
- Oligotex mRNA isolation kit which includes 2 × binding buffer, oligotex suspension, wash, and elution buffers) (Qiagen)
- 75 μl of total RNA (DNase-treated)
- Chloroform; isopropanol; 70% ethanol
- RNAse-free water

Method

1 Add 100 μl of 2 × binding buffer directly to the 75 μl of aqueous phase from the DNase treated total RNA[c].

Protocol 3 continued

2 Add 6 μl of Oligotex suspension, mix well, and incubate at 65 °C for 3 min.
3 Incubate the tube at room temperature for 15 min, mixing gently every 3 min, and pre-warm the elution buffer to 75 °C.
4 Spin at 14 500 g for 3 min at room temperature and discard the supernatant.[d]
5 Wash the Oligotex twice in 300 μl of wash buffer, each time flicking the tube to resuspend, and centrifuging for 3 min. Remove the wash buffer carefully.
6 Elute the mRNA from the Oligotex with 6 μl of pre-heated elution buffer, by pipetting the Oligotex up and down, vortexing and incubating for 2 min at 75 °C.
7 Spin at 14 500 g for 3 min at room temperature. Remove the supernatant to a fresh tube and repeat. Combine the supernatants so that the poly(A)$^+$ RNA is dissolved in 12 μl of elution buffer.

[a] This is a modification of the protocol supplied by Qiagen, manufacturers of the Oligotex mRNA isolation kit.

[b] The protocol has been successfully applied to between 10^4 and 10^7 lymphocytes (approximately 100 ng and 100 μg of total RNA, respectively). mRNA typically forms 1–2% of total RNA. For between 10^7 and 10^8 cells (or 100 μg and 1 mg of total RNA) we recommend following the protocol supplied by the manufacturer of Oligotex.

[c] Be sure to dissolve any precipitate present in the 2 × binding buffer before use.

[d] The Oligotex pellet is fragile and it is safer not to attempt to take too much of the supernatant. We also suggest that supernatants and washes are respun to ensure material is not lost.

Protocol 4

Preparation of poly(A)$^+$ RNA (mRNA) from $>10^8$ cells[a]

Equipment and reagents

- Disposable columns (Biorad Polyprep chromatography columns cat. no. 731–1550)
- 1 ml of total RNA (DNase-treated)
- 0.5 M NaCl (RNase-free)
- High Salt Buffer (400 mM NaCl, 10 mM Tris pH 7.4, 1 mM ethylenediamine tetraacetic acid (EDTA))
- Medium Salt Buffer (200 mM NaCl, 10 mM Tris pH 7.4, 1 mM EDTA)
- Low Salt Elution Buffer (10 mM Tris pH 7.4, 1 mM EDTA)
- Oligo(dT) cellulose (Type 3, Collaborative Research)
- RNAse-free water

Method[b]

1 Prepare total RNA from cells or tissues using Protocol 1 or another suitable procedure. Resuspend in 1 ml of RNase free water.
2 Add 1 ml of total RNA to 5 ml 0.5 M NaCl in a 50 ml Falcon tube.
3 Add 0.5–1.0 ml (for lower quantities, still use about 0.5 ml) packed volume of equilibrated oligo(dT) cellulose.

Protocol 4 continued

4 Incubate for 1 h at room temperature, turning slowly on a rotating wheel.

5 Spin down the cellulose at 1500 g for 30 s. Pour off the supernatant.

6 Resuspend the oligo(dT) cellulose in 10 ml of High Salt Buffer and centrifuge it as before. Repeat twice.

7 Resuspend the cellulose in 10 ml of High Salt Buffer, and pipette or pour it into a pre-washed disposable column. Allow the buffer to drip to waste, and wash with 10 ml of High Salt Buffer.

8 Wash the column with 3 ml of Medium Salt Buffer. Heat the Low Salt Elution Buffer to 50 °C.

9 Elute the poly(A)$^+$ RNA with 1.5 ml of Low Salt Elution Buffer at 50 °C. Remember to collect this material!

10 Prepare a 1:10 dilution of the eluted material and measure the concentration on a spectrophotometer at 260/280.[d]

11 Add ethanol to 70% and store the poly(A)$^+$ RNA in aliquots at −80 °C. If the mRNA is very dilute at this stage it is better to ethanol precipitate and resuspend it in a smaller volume (for example 0.5 μg μl^{-1}).[e]

[a] This protocol is offered as a cheaper and effective alternative to the Qiagen Oligotex kit. It is preferable to begin with at least 10^8 cells (or approximately 1 mg total RNA). For between 10^7 and 10^8 cells (or 100 μg and 1 mg of total RNA) we recommend using the Oligotex kit, following the protocol supplied by the manufacturer

[b] Volumes assume cell number of 2–5 × 10^8 cells (approximately one T175 culture flask). Scale up or down accordingly.

[c] To facilitate binding to the oligo(dT) cellulose.

[d] Approximately 50% of the RNA will still be rRNA after one oligo(dT) selection. Performing a second oligo(dT) selection can increase mRNA purity and help eliminate DNA contamination.

[e] Store the mRNA in 70% ethanol, but without salt, which should only be added at the time of precipitation.

4.3 Generation of double-stranded cDNA

Because it is necessary to create cDNA, which can be digested by a restriction enzyme, double-stranded cDNA must be prepared. The primer used to generate the first strand cDNA is normally an oligo(dT) oligonucleotide of 18–20 bases. At this point it is worth considering using a modified oligo(dT) primer which contains a *Dpn*II site to increase representation of cDNA fragments at a later stage. In this case it is crucial to anchor the primer to the mRNA at its 3′ end by including two degenerate bases, for example using a primer such as 5′-$(N)_6GATC(T)_{18-30}N_1N_2$-3′, where N_1 can be A, C, or G, and N_2, A, C, G, or T. This is to ensure that all the 3′ terminal gene fragments are of the same length and will exhibit the same hybridization kinetics. Another useful modification is to synthesize the first strand of cDNA with a primer compatible with future RACE or cDNA library strategies to be used in the secondary screening process for identifying full length cDNA clones. This is particularly helpful if no appropriate cDNA library is available. The CLONTECH 5′-3′ RACE kit is a useful example of this type of approach. In this case we would caution against engineering a *Dpn*II site in the primer at the 5′ end of the gene. This will create a range of 5′-derived fragments of different sizes depending on the efficiency of RT. The subsequent behaviour of these fragments in the hybridizations may not be compatible with cDNA RDA.

Protocol 5

Preparation of double-stranded cDNA from <10^7 cells

Equipment and reagents

- Refrigerated microcentrifuge
- 12 μl of poly(A)$^+$ RNA (from *Protocol 3*)
- 5 × First Strand Buffer (e.g. SuperScript II buffer, Life Technologies)
- 5 × RT2 Second Strand Buffer (1 ml of 5 × RT2 is prepared by mixing 100 μl of 1 M Tris–HCl (pH 7.5), 500 μl of 1 M KCl, 25 μl of 1 M $MgCl_2$, 50 μl of 1 M $(NH_4)_2SO_4$, 50 μl of 1 M DTT, 50 μl of 5 mg ml^{-1} bovine serum albumin (BSA), and 225 μl of water)
- dNTP mixture (20 mM each of dATP, dCTP, dGTP, TTP), Oligo(dT) primer (Promega; 50 ng μl^{-1}), RNase inhibitor (20–40 U μl^{-1}), 15 mM βNAD (Boehringer Mannheim)
- RT (e.g. SuperScript II, Life Technologies, 200 U μl^{-1}), *Escherichia coli* DNA ligase (NEB, 5 U μl^{-1}), RNase H (NEB, 2.5 U μl^{-1}), *E. coli* DNA polymerase (NEB, 10 U μl^{-1})
- RNAse-free water

First strand cDNA synthesis

1. Isolate mRNA in 12 μl RNase-free water by *Protocol 3*. Respin the mRNA at 14 500 g for 4 min at 4 °C to pellet any carry-over Oligotex. Move 10 μl into a fresh 0.5 ml microcentrifuge tube to be designated the +RT sample. Add 8 μl of RNase-free water to the first tube; this will form the −RT sample.[a]
2. Incubate the mRNA at 70 °C for 5 min,[b] then place on ice.
3. To each RNA sample add 4 μl of 5 × first strand buffer, 1 μl of 20 mM dNTPs, 2 μl of 0.1 M DTT, 1 μl of 50 ng μl^{-1} oligo(dT) primer, 0.5 μl of RNase inhibitor and 0.5 μl of RNase-free water (1.5 μl to the −RT sample). Add 1 μl of RT to the '+RT' tube and incubate at 26 °C for 8 min followed by 41 °C for 1 h 30 min.[c]

Second strand cDNA synthesis

1. Prepare a premix consisting of 1.8 μl of double deionized H_2O, 6 μl of 5 × RT2 buffer, 0.5 μl 15 mM βNAD, 0.4 μl of *E. coli* DNA ligase, 0.3 μl of RNase H, and 1 μl of *E. coli* DNA polymerase.[d]
2. Add 10 μl of the premix to each tube of first strand synthesis reaction. Incubate at 15 °C for 2 h, then 22 °C for 1 h.[e]
3. Denature the enzymes used to synthesize cDNA by heating to 70 °C for 10 min, then place the sample on ice.

Protocol 5 continued

4 Precipitate the cDNA by adding 1 μl of glycogen carrier (5 mg ml^{-1}), 3 μl of 3 M sodium acetate and 90 μl of 100% ethanol, then cool to −20 °C for 20 min.

5 Centrifuge at 14 000 r.p.m. for 30—60 min at 4 °C. Wash the pellet with 70% ethanol and resuspend it in 17 μl of water, and proceed directly to digest the cDNA with *Dpn*II.

[a] Only 20% of the material is used as the negative control to conserve RNA. It is important not to omit this control as it is the only way to assess potential contamination by genomic DNA.

[b] To disrupt secondary structure.

[c] Oligo(dT) binds to the 3′ poly(A) tail of the mRNA and primes first strand synthesis. Conditions need to be optimal to obtain the 5′ ends of long messages. This can be monitored by including ^{32}P-dCTP in a pilot reaction (or an aliquot), running it on an alkaline agarose gel and exposing it to film (18).

[d] During second strand synthesis, RNase H nicks the RNA in the DNA:RNA hybrid product of first strand cDNA synthesis. DNA polymerase extends the resulting 5′ ends of RNA, using the first strand as a template. Finally, after all the RNA is degraded, the multiple short DNA second strands are ligated together by *E. coli* DNA ligase, using βNAD as a cofactor.

[e] At this point the reaction can be left overnight at 4 °C.

Protocol 6

Preparation of double-stranded cDNA from >10^7 cells

Equipment and reagents

- See *Protocol 5*
- 100 mM sodium pyrophosphate
- 1–5 μg of poly$(A)^+$ RNA (from *Protocols 3* or *4*)
- TE (10 mM Tris pH 7.4; 1 mM EDTA)
- P/C/I; C/I (see *Protocol 2*)

First strand cDNA synthesis

1 Begin with 1–5 μg of mRNA dissolved in 20 μl of RNase-free water.[a] Heat it to 70 °C for 10 min. Chill on ice.

2 Add in order to the mRNA: 8 μl of 5 × first strand buffer, 0.5 μl of 100 mM DTT, 0.7 μl of RNase inhibitor, 2 μl of oligo(dT) primer, 4.0 μl of dNTPs, 1.6 μl of 100 mM sodium pyrophosphate, and 3.3 μl of RT. Incubate at 26 °C for 8 min followed by 41 °C for 1 h 30 min.

Second strand cDNA synthesis

1 Dilute 40 μl of 5 × RT2 with 112.5 μl of water and add it to the completed first strand synthesis reaction.

2 Add in order: 2.0 μl of βNAD, 0.4 μl of *E Coli* DNA ligase, 4.0 μl of *E Coli* DNA polymerase and 1.1 μl of RNase H. Incubate for 2 h at 15 °C, then for 1 h at 22 °C.

Protocol 6 continued

3 Add 50 μl of water. Extract twice with P/C/I, and once with C/I. Add 5 μg of glycogen carrier, 65 μl of 10 M NH_4OAc, and 800 μl of 100% ethanol. Precipitate for 20 min on ice and spin at 14 500 g for 15 min at 4 °C. Wash the pellet with 500 μl of 70% ethanol, and resuspend it in 30 μl of TE.

4 Run 3 μl on a 0.7% agarose gel.[b]

[a] A negative control (without RT) can be performed to assess potential contamination by genomic DNA. However, when 1–5 μg of carefully isolated poly(A)$^+$ RNA serve as the template, it is unlikely that DNA will form a very significant portion of the resulting cDNA.

[b] A smear from approximately 8–10 kb down to 0.2 kb should be expected. Yield should be assessed, and is usually approximately 50%. (Taking into account double strandedness of cDNA, this is 1 μg of cDNA per microgram of poly(A) RNA.)

4.4 Preparation of representations

Representations form the basis of cDNA RDA subtractions. To prepare them, double-stranded cDNA is cut with *Dpn*II, unequal adaptors are ligated on and a fill-in reaction is performed prior to amplification by PCR.

Protocol 7

*Dpn*II digestion of double-stranded cDNA

Equipment and reagents

- Double-stranded cDNA (from *Protocols* 5 or 6)
- P/C/I; C/I (see *Protocol* 2)
- *Dpn*II (NEB, 10 U μl^{-1}), 10 × DpnII buffer (NEB)

A From <10^7 cells, or <1 mg of cDNA

1 Digest all 17 μl of the cDNA from *Protocol* 5 by adding 2 μl of 10 × *Dpn*II buffer and 1 μl of *Dpn*II. Incubate for 1–2 h at 37 °C.

2 Denature the *Dpn*II for 20 min at 65 °C. Chill on ice and spin briefly to collect the sample at the bottom of the tube.

B From >1 mg of cDNA

1 Digest 1–2 μg of double-stranded cDNA from *Protocol* 6 for 2 h at 37 °C, using 2 μl of *Dpn*II and 10 μl of 10 × *Dpn*II buffer, in a total volume of 100 μl.

2 Extract the reactions with P/C/I, then with C/I. Add 50 μl of 10 M NH_4OAc, 650 μl of 100% ethanol and precipitate on ice for 20 min.[a]

3 Centrifuge the sample at 14 500 g for 15 min at 4 °C. Wash the pellet with 70% ethanol. Air-dry the pellet and resuspend it in 20 μl of TE.

[a] 10 μg of glycogen carrier can be used if required.

The oligonucleotides used for the ligations should be desalted and preferably high-performance liquid chromatography (HPLC) purified. Many manufacturers of oligonucleotides supply them already deprotected and desalted. Alternatively, a procedure for desalting can be found in Lisitsyn and Wigler (3). During the ligation, the 12-mer acts as a splice to permit ligation of the 24-mer to the 5′ ends of the fragments. Because the oligonucleotides are not phosphorylated, only the 24-mer (R-24) is covalently linked to the 5′ phosphate group of the *Dpn*II site from the digested cDNA, while the 12-mer remains annealed, but unligated. The fill-in is preferable to ligation of phosphorylated adaptors in generating the representation as it prevents the formation of oligomers of adaptors on individual fragments.

Protocol 8

Ligation of *Dpn*II digested cDNA to R-adaptors

Equipment and reagents

- PCR machine
- Ligation water bath
- T4 DNA ligase (NEB, 400 U μl^{-1}), 10 × ligase buffer (500 mM Tris–HCl (pH 7.8); 100 mM $MgCl_2$; 100 mM DTT; 10 mM adenosine triphosphate (ATP)); 10 mM ATP
- *Dpn*II digested double-stranded cDNA (from *Protocol 7*)
- Oligonucleotides R-12 (5′-GATCTGCGGTGA-3′; 1 mg ml^{-1}) and R-24 (5′-AGCACTCTCCAGCCTCTCACCGCA-3′; 2 mg ml^{-1})[a]

A From <10^7 cells, or <1 mg of cDNA

1. Ligate the entire 20 μl of the cDNA reaction from *Protocol 7A* to R-adaptors by mixing 31 μl of water, 4 μl of 10 × ligase buffer,[b] 2 μl of 10 mM ATP, 1 μl of R-24, and 1 μl of R-12 in a 0.5 ml microcentrifuge tube. Add 39 μl of the mixture to 20 μl of digested cDNA.
2. Anneal the oligonucleotides by heating the reaction to 50 °C for 1 min, then cooling to 10 °C at 1 °C min^{-1}.[c]
3. Add 1 μl of T4 DNA ligase, mix, and incubate overnight at 14 °C.

B From >1 μg of cDNA

1. Ligate the digested cDNA to R-adaptors by mixing 12 μl (approximately 1.2 μg) cut cDNA with 4 μl of R-24, 4 μl of R-12, 6 μl of 10 × ligase buffer and 32 μl of water in a 0.5 ml microcentrifuge tube.
2. Anneal the oligonucleotides by heating the reaction to 50 °C for 1 min, then cooling to 10 °C at 1 °C min^{-1}.[c]
3. Add 2 μl of T4 DNA ligase, mix, and incubate overnight at 14 °C.

[a] The oligonucleotides should be prepared on a large scale (1 μM), desalted, preferably HPLC purified, and resuspended in water.

[b] The reduced volume of 10 × ligation buffer added takes into account the salt concentration of the *Dpn*II buffer, and requires additional ATP. Alternatively, a 10 × ligation buffer containing reduced salt and increased ATP can be prepared especially for this ligation.

[c] It is convenient to program a PCR machine to perform the gradual cooling.

We strongly recommend that a pilot reaction be performed for each representation to be generated and for the corresponding −RT control. This is to establish the optimum amplification conditions, to judge quantities, and to monitor genomic DNA contamination before the entire stock of material is committed.

Protocol 9
Generation of representations

Equipment and reagents

- PCR machine with heated lid
- R-ligated *Dpn*II fragments of cDNA (from *Protocol 8*)
- Oligonucleotide primer R-24 (5′-AGCACTCTCCAGCCTCTCACCGCA-3′; 2 mg ml^{-1})[a]
- *Taq* DNA polymerase (e.g. Gibco BRL, 5 U μl^{-1}); 5 × PCR buffer (335 mM Tris–HCl, pH 8.9 at 25 °C, 20 mM $MgCl_2$, 80 mM $(NH_4)_2SO_4$, 166 μg ml^{-1} BSA); 4 mM dNTPs (4 mM each dATP, dCTP, dGTP, and TTP)
- P/C/I; C/I (see *Protocol 2*)

A Pilot reactions

1 For each pilot reaction to be performed, prepare a 0.5 ml microcentrifuge tube containing 140 μl of water, 40 μl of 5 × PCR buffer, 16 μl of 4 mM dNTPs, and 1 μl of R-24 primer.

2a If you prepared your ligation on a small scale (*Protocol 8A*), add 2 μl of the ligation reaction to the PCR mix. Place in a PCR machine equipped with a heated lid or overlay with mineral oil. Proceed to step 3.

2b If you prepared your ligation on a large scale (*Protocol 8B*), dilute the ligation to 10 ng μl^{-1} by adding 140 μl of TE, then add 2 μl of the diluted ligation reaction to the PCR mix. Place in a PCR machine equipped with a heated lid or overlay with mineral oil. Proceed to step 3.

3 Incubate at 72 °C for 3 min.[b]

4 Pause and add 1 μl of *Taq* DNA polymerase and incubate at 72 °C for 5 min.[c]

5 Cycle at 95 °C for 1 min, then at 72 °C for 3 min for various numbers of cycles (see step 6), finishing with a final extension at 72 °C for 10 min.[d]

6a If you prepared your ligation on a small scale (*Protocol 8A*), pause the reaction after 18, 20, 22, and 24 cycles and remove 10 μl aliquots of the product. Proceed to step 7.

6b If you prepared your ligation on a large scale (*Protocol 8B*), pause the reaction after 17, 18, 19, and 20 cycles and remove 10 μl aliquots of the product. Proceed to step 7.

7 To each aliquot add 1 μl of 10 × agarose gel loading buffer and load on to a 1.5% agarose gel containing ethidium bromide, together with appropriate DNA concentration standards and molecular weight markers.

Protocol 9 continued

B Full-scale generation of representations

1. Study the gel (*Protocol 9A*, step 7), and select a cycle number that produces a smear ranging in size from 0.2 – 1.5 kb, and containing approximately 0.5 μg of DNA in the 10 μl aliquot.
2. Set up eight reactions (*Protocol 9A*, step 1) for each sample intended for use as driver, and two reactions for each sample intended to serve solely as tester.
3. Proceed through *Protocol 9A*, performing the appropriate number of cycles in step 6, followed by a final extension at 72 °C for 10 min.
4. Combine the eight reactions into two 1.5 ml microcentrifuge tubes (four reactions in each) and extract each with 700 μl of P/C/I, then with 700 μl of C/I. Add 75 μl of 3 M sodium acetate (pH 5.3), 750 μl of isopropanol, and precipitate the DNA on ice for 20 min.[e]
5. Centrifuge at 14 500 g for 15 min at 4 °C. Wash the pellet with 0.5 ml of 70% ethanol, dry, and resuspend each of the pellets in 100 μl of TE to give a concentration of 0.5 mg ml^{-1}.[f]
6. Combine the products and run 1 μl on a 1.5% agarose gel to check concentration and quality (*Protocol 9A* step 7).

[a] The oligonucleotides should be prepared on a large scale (1 μM), desalted, preferably HPLC purified, and resuspended in water.

[b] During this period the 12-mer (R-12) dissociates, freeing the 3′ ends for subsequent fill-in.

[c] To fill in the ends complementary to the 24-mer adaptors

[d] Both primer annealing and extension occur at 72 °C.

[e] If there are only two reactions, follow the protocol in a single tube, using half the indicated volumes.

[f] Resuspending the pellet in 25–30 μl of TE for each original reaction it contains (up to four) usually provides the correct concentration.

It is important that concentrations are accurately determined, so the combined representation should again be run next to a previously prepared cut standard to allow comparison. Concentration standards can be prepared by digesting genomic DNA of a known concentration (sheared to average length of approximately 20 kb) with a restriction enzyme that has a 4-bp recognition site, such as *Sau*3A. Dilutions corresponding to 0.1, 0.5, and 1 μg of the standard should be loaded along with the PCR product on to a 1.5% agarose gel containing ethidium bromide. After a short (2–3 cm) electrophoresis, yield should be assessed on an ultraviolet transilluminator, and should ideally be approximately 0.5 μg from the 10 μl aliquot. Reliance on spectrophotometer readings is inadequate. Once a successful representation has been generated it can be used as the standard for other experiments.

The pilot reactions should show a smear in the +RT lane while the −RT lane should be clear or very faint. Material amplifying from the −RT reaction arises from genomic contamination in the RNA preparation. Remember that the −RT lane contains only 20% of the template in the +RT lane.

The gel should be run further to assess the size distribution of products. A smear ranging in size from 1.5 to 0.2 kb should be obtained. Discrete bands may be visible, representing amplified fragments of abundant transcripts, and these will be characteristic of a particular sample. Where closely matched, tester and driver samples to be subtracted will not appear different at this stage. A product that matches the correct criteria of yield, size distribution, and a clean −RT control, indicates the number of cycles necessary for preparing the representation. Underamplification will not provide sufficient material for the subtraction, while overamplification will bias the population and reduce the average fragment size. Should more than 24 cycles be required this probably signals a problem with the mRNA isolation or cDNA synthesis, and it is better to begin again. If only two rounds of subtraction are to be performed, a minimum of 20 μg of driver and 1 μg tester will be required. However, it is wise to prepare a significantly larger quantity to allow for the preparation of representation blots and for extra subtractions and alterations in stringency, which may be required at a later stage.

Usually, a representation will be used as both a tester and a driver, in a reciprocal subtraction.

4.4 Preparation of driver and tester

Driver is prepared simply by digesting away the R-adaptors used to generate the representation. It requires no further purification. The tester is prepared by cutting the representation and ligating on a new set of unequal adaptors (J-adaptors), which will create the template from which a primer binding site will be generated after hybridization. To achieve this efficiently, the majority of R-adaptors should be removed either by gel purification or by spin column preparation, to prevent their reannealing to the fragments during the ligation to J-adaptors.

Protocol 10

Generation of driver

Equipment and reagents

- Representative amplicons (0.5 μg μl^{-1}; from *Protocol 9*)
- P/C/I; C/I (see *Protocol 2*)
- *Dpn*II (NEB, 10 U μl^{-1}), 10 × DpnII buffer (NEB)

Method

1 Digest 90 μg (180 μl) of each representation with *Dpn*II. Mix 180 μl of the representation DNA with 340 μl of water, 60 μl of 10 × *Dpn*II buffer, and 20 μl of *Dpn*II. Incubate for 2 h at 37 °C.

2 Extract with equal volumes of P/C/I, then C/I. Add 60 μl of 3 M sodium acetate (pH 5.3) and 600 μl of isopropanol, and precipitate on ice for 20 min.

Protocol 10 continued

3 Centrifuge the sample at 14 500 g for 15 min at 4 °C. Wash the pellet with 70% ethanol. Air-dry the pellet and resuspend the digested representation in 180 μl of TE.

4 Assess the concentration by running 1 μl on a 1.5% gel with standards, then dilute it if necessary with TE to 0.5 mg ml^{-1}. This is the cut DRIVER.

[a] This will provide sufficient driver for up to nine subtractions.

Protocol 11

Generation of tester

Equipment and reagents

- Representative amplicons (0.5 μg μl^{-1}, from *Protocols 9* or *10*)
- *Dpn*II (NEB, 10 U μl^{-1}), 10 × DpnII buffer (NEB)
- QIAquick PCR purification kit, including spin columns; binding buffer PB, and wash buffer PE, (Qiagen)
- J-12 oligonucleotide (5′-GATCTGTTCATG-3′; 1 mg ml^{-1}), J-24 oligonucleotide (5′-ACCGACGTCGACTATCCATGAACA-3′; 2 mg ml^{-1})
- T4 DNA ligase (NEB, 400 U μl^{-1}), 10 × ligase buffer (500 mM Tris–HCl (pH 7.8); 100 mM $MgCl_2$; 100 mM DTT; 10 mM ATP)

A Purification of representation

1a If a representation is to be serve as both tester and driver in a reciprocal subtraction, follow *Protocol 10*, remove 20 μl of the final digested product to a separate tube and proceed to Protocol 11, step 4.

1b If a representation is to be used only as a tester, digest 10 μg (20 μl) of each representation (from *Protocol 9*) with *Dpn*II. Mix 20 μl of the representation DNA with 68 μl of water, 10 μl of 10 × *Dpn*II buffer, and 2 μl of *Dpn*II. Incubate for 2 h at 37 °C.

2 Extract with equal volumes of P/C/I, then C/I. Add 10 μl of 3 M sodium acetate (pH 5.3), 250 μl of 100% ethanol, and precipitate at −20 °C for 20 min.

3 Centrifuge the sample at 14 500 g for 15 min at 4°C. Wash the pellet with 70% ethanol. Air-dry the pellet and resuspend the digested representation in 20 μl of TE.

4 Purify the representation away from the digested R-adaptors by using a spin purification column such as QIAquick spin columns.[a,b]

5 Place a QIAquick spin column in a 2 ml collection tube. Add 100 μl of Buffer PB (Qiagen) to 20 μl of digested representation and apply the mixture to the column and allow it to stand for 2 min.

6 Centrifuge the column at 14 500 g for 30–60 s at room temperature. Discard the flow through.

7 Wash the column with 0.75 ml of Buffer PE (Qiagen), centrifuging again for 30–60 s. Discard the flow through and centrifuge for an additional 1 min to remove any residual buffer.

Protocol 11 continued

8 Elute the DNA by the addition of 50 μl of TE to the centre of the column. Allow the column to stand for 2 min, then centrifuge as before for 1 min.

9 Repeat the elution and combine the eluates, to give 100 μl of cut purified representation in TE.

10 Estimate the concentration of the DNA and the efficiency of adaptor removal by running a 5 μl aliquot next to standards on a 1.5% agarose gel.[c,d]

B Ligation to J-adaptors

1 Ligate the cut, purified representation to J-adaptors in a 0.5 ml microcentrifuge tube by mixing 1 μg of spin column-purified DNA (usually 10–20 μl) with 3 μl of 10 × ligase buffer, 2 μl of 1 mg ml^{-1} J-12 oligo, and 2 μl of 2 mg ml^{-1} J-24 oligo. Add water to a total volume of 29 μl.

2 Anneal the oligos in a PCR machine by heating to 50 °C for 1 min, then cooling to 10 °C at 1 °C min^{-1}.

3 Add 1 μl of T4 DNA ligase, and incubate the reaction overnight at 12 °C.

4 Dilute the ligation to 10 ng μl^{-1} by adding 70 μl TE. This generates the J-Ligated TESTER.

[a] Adaptors can also be removed by gel purification or by other related approaches.

[b] The protocol provided is modified, with permission, from the protocol provided by Qiagen.

[c] A small quantity of R-adaptors may still be visible, but these should not significantly affect the following ligation because J-adaptors are added in a large excess.

[d] It is normal for 5–10 μl to correspond to approximately 0.5 μg of cut, purified representation.

Protocol 12

Subtractive hybridization

Equipment and reagents

- PCR machine
- Cut Driver representation (0.5 μg μl^{-1}, from *Protocol 10*)
- J-ligated Tester representation (10 ng μl^{-1}, from *Protocol 11*)
- 10 M ammonium acetate (NH_4OAc); EE × 3 buffer (30 mM EPPS (Sigma), pH 8.0 at 20 °C; 3 mM EDTA); 5 M NaCl

Method

1 Mix 20 μl (10 μg) of digested DRIVER representation with 10 μl of diluted, J-ligated TESTER representation (0.1 μg) in a 0.5 ml microcentrifuge tube, a driver/tester ratio of 100:1.

2 Add 70 μl of water and extract with 100 μl of P/C/I, followed by 100 μl of C/I. Add 25 μl of 10 M NH_4OAc, 320 μl of 100% ethanol, and precipitate at −70 °C for 10 min.

Protocol 12 continued

3 Incubate the tube at 37 °C for 1 min,[a] then centrifuge at 14 000 r.p.m. for 15 min at 4 °C. Wash the pellet twice with 70% ethanol, each time spinning at 14 500 g for 2 min.[b]

4 Air dry the pellet, and resuspend it very thoroughly in 4 μl of EE × 3 buffer by pipetting for at least 2 min, then warming to 37 °C for 5 min, vortexing, and spinning to the bottom of the tube.

5 Overlay the solution with 35 μl of mineral oil, place the tube in a PCR machine and denature the DNA for 5 min at 98 °C. Cool the block to 67 °C, and immediately add 1 μl of 5 M NaCl directly into the DNA.

6 Incubate the hybridization at 67 °C for 20 h.[c]

7 Remove as much oil as possible from the completed hybridization[d] and dilute the DNA stepwise in 200 μl of TE. The DNA can be very viscous at this point and should be diluted first by adding 10 μl TE and repeatedly pipetting, followed by a further 25 μl TE with more pipetting, and finally made up to 200 μl and vortexed.

8 Store the hybridized DNA at −20 °C, or use it directly to generate the first difference product.

[a] To reduce salt precipitation.

[b] If the pellet is very difficult to see after the first wash, omit the second wash.

[c] To allow complete hybridization.

[d] Residual oil in the dilution does not impair the subsequent amplification.

4.4 Subtractive hybridization

During the subtraction, the driver and tester are mixed and allowed to anneal under optimal conditions of salt concentration and temperature. The ratio of driver to tester determines the stringency of the subtraction, and a balance is achieved between the formation of unamplifiable driver/tester hybrids, which removes common material, and the desire to retain amplifiable tester/tester hybrids forming at very low concentrations. Complete denaturation of both driver and tester is essential at this stage as any tester DNA that fails to denature will remain amplifiable, regardless of its abundance in the driver.

4.5 Generation of a first difference product (DP1)

The generation of a difference product is a three-step process. The first stage PCR performs two critical functions. First, the initial fill-in generates the binding site for the primer (J-24) binding, and is the reason why only tester/tester hybrids amplify exponentially. Without the fill-in reaction, the material will not be amplifiable. The amplification achieved after eleven cycles of PCR also guards against potential loss of rare but genuine products during the precipitation steps prior to mung bean nuclease (MBN) treatment. The MBN digestion removes all single-stranded nucleic acids, including the melted and unamplifiable driver, linear products of the driver/tester hybrids, and any tRNA used as a carrier in the precipitation. The final PCR amplifies the tester/tester hybrids to produce sufficient material to proceed to the next stage of the procedure.

Protocol 13

Generation of the first difference product (DP1)

Equipment and reagents

- PCR machine with heated lid
- Hybridized driver and tester (200 μl, from *Protocol 12*)
- *Taq* DNA polymerase (e.g. Gibco BRL, 5 U μl^{-1}); 5 × PCR buffer (335 mM Tris–HCl, pH 8.9 at 25 °C, 20 mM $MgCl_2$, 80 mM $(NH_4)_2SO_4$, 166 $\mu g\, ml^{-1}$ BSA); 4 mM dNTPs (4 mM each dATP, dCTP, dGTP and TTP); tRNA carrier (10 $\mu g\, \mu l^{-1}$)
- Oligonucleotide primer J-24 (5′-AGCACTCTCCAGCCTCTCACCGCA-3′; 2 mg ml^{-1})[a]
- Mung Bean Nuclease (MBN) (NEB; 10 U μl^{-1}); 10 × MBN buffer (NEB); 50 mM Tris–HCl (pH 8.9)

A Initial amplification

1. For each subtraction set up two PCR reactions in 0.5 ml microcentrifuge tubes. For each reaction, mix in order: 122 μl of water, 40 μl of 5 × PCR buffer, and 16 μl of 4 mM dNTPs. Finally, add 20 μl of the diluted hybridization mix.
2. Place in a PCR machine equipped with a heated lid or overlay with mineral oil.
3. Incubate at 72 °C for 3 min.[b]
4. Pause the machine and add 1 μl of Taq DNA polymerase to each reaction, and incubate at 72 °C for 5 min.[c]
5. Just before the 5 min incubation is finished, pause the machine again, and add 1 μl of 2 mg ml^{-1} J-24 primer.
6. Cycle at 95 °C for 1 min, then at 70 °C for 3 min for 11 cycles, finishing with a final extension at 72 °C for 10 min.[d]
7. Combine the two reactions into a single 1.5 ml microcentrifuge tube. Extract with 400 μl of P/C/I, and then C/I. Add 10 μl of tRNA carrier, 40 μl of 3 M sodium acetate (pH 5.3), 400 μl of isopropanol, and precipitate on ice for 20 min.
8. Centrifuge at 14 500 *g* for 15 min at 4°C, and wash the pellet with 70% ethanol. Resuspend the pellet in 20 μl of TE.

B MBN digestion

1. Transfer the 20 μl of DNA from *Protocol 13A*, step 8 to a 0.5 ml microcentrifuge tube and add 4 μl of 10 × MBN buffer, 14 μl of water, and 2 μl of MBN.
2. Incubate the tube in a PCR machine at 30 °C for 35 min.
3. Stop the reaction by adding 160 μl of 50 mM Tris–HCl (pH 8.9), and incubating at 98 °C for 5 min. Place the tube containing the denatured, MBN-treated DNA on ice.

Protocol 13 continued

C Amplification to DP1

1. For each subtraction set up one PCR reaction in a 0.5 ml microcentrifuge tube by combining 122 μl of water, 40 μl of 5 × PCR buffer, 16 μl of 4 mM dNTPs, and 1 μl of 2 mg ml^{-1} J-24 oligo. Place the tubes on ice.
2. On ice, add 20 μl of the MBN-treated DNA.
3. Place the tubes in a PCR machine and incubate at 95 °C for 1 min. Cool the block to 80 °C and add 1 μl of Taq DNA polymerase.
4. Perform 18 cycles each of 95 °C for 1 min then 70 °C for 3 min, with a final extension of 72 °C for 10 min. Cool the reactions to room temperature.
5. Check a 10 μl sample on a 1.5% agarose gel next to standards to confirm the presence of products and to estimate the concentration.
6. Extract with 200 μl of P/C/I, and then C/I. Add 20 μl of 3 M sodium acetate (pH 5.3), 200 μl of isopropanol, and precipitate the products on ice for 20 min.
7. Centrifuge at 14 500 g for 15 min at 4 °C, and wash the pellet with 70% ethanol. Resuspend the pellet at 0.5 μg μl^{-1} in TE[e] (approximately 25 μl, depending on the estimated yield—determined in step 5). This is the first difference product (DP1)

[a] The oligonucleotides should be prepared on a large scale (1 μM), desalted, preferably HPLC purified, and resuspended in water.

[b] During this period the 12-mer (R-12) dissociates, freeing the 3′ ends for subsequent fill-in.

[c] To fill in the ends complementary to the 24-mer oligo.

[d] Note that J-primers have a preferred annealing temperature of 70 °C.

[e] Resuspending the pellet in 25–30 μl of TE for each original reaction it contains (up to four) usually provides the correct concentration.

To monitor the success of MBN treatment, a control can be performed in which the MBN is omitted. Eleven-cycle products from treated and untreated reactions should then be compared on a 1.5% agarose gel. Most of the ethidium bromide staining should be eliminated from the MBN treated sample. The yield of the final reaction should be 10—12 μg, and the product may appear enriched for certain bands. If the yield is significantly (more than twofold) lower, it is likely that a problem has arisen, either during the PCR, or at an earlier stage in the procedure. In such cases, it is advisable to return and perform careful controls to determine the source of the difficulty (see Section 5 on common problems).

4.5 Generation of a second difference product (DP2)

In generating the second difference product, the J-adaptors are removed from DP1, new N-adaptors are ligated to form a second round tester, and the subtractive hybridization and amplification are repeated. During the ligation step, dilution of the digested difference product ensures that religation of the free J oligos still present will be insignificant owing to very high levels of N oligos in the reaction. Because the enrichment of differences has already been achieved at DP1, the ratio of driver/tester in the hybridization is more stringent (800:1), driving out the majority of unwanted material. The increased kinetic enrichment of tester/tester hybrids in the second subtraction ensures that the DP2 will usually consist of distinct bands containing predominantly differences between the populations.

Protocol 14

Change of adaptors on a difference product

Equipment and reagents

- Difference product (0.5 μg μl^{-1}, from *Protocol 13*)
- N-12 oligonucleotide (5′-GATCTTCCCTCG-3′; 1 mg ml^{-1}), N-24 oligonucleotide (5′-AGGCAACTGTGCTATCCGAGGGAA-3′; 2 mg ml^{-1})
- *Dpn*II (NEB, 10 U μl^{-1}), 10 × DpnII buffer (NEB); glycogen carrier (10 μg μl^{-1}, Boehringer Mannheim)
- T4 DNA ligase (NEB, 400 U μl^{-1}), 10 × ligase buffer (500 mM Tris–HCl (pH 7.8); 100 mM $MgCl_2$; 100 mM DTT; 10 mM ATP)

A Removal of adaptors

1. Digest 2 μg (4 μl) of the difference product (from *Protocol 13*) with DpnII. In a 1.5 ml microcentrifuge tube, mix 4 μl of the DP1 DNA with 84 μl of water, 10 μl of 10 × DpnII buffer, and 2 μl of DpnII. Incubate for 2 h at 37 °C.
2. Extract with equal volumes of P/C/I, then C/I. Add 1 μl of glycogen carrier, 10 μl of 3 M sodium acetate (pH 5.3), 250 μl of 100% ethanol, and precipitate at −20 °C for 20 min.
3. Centrifuge the sample at 14 500 g for 15 min at 4 °C. Wash the pellet with 70% ethanol. Air-dry the pellet and resuspend the digested representation in 20 μl of TE (100 ng μl^{-1}).

B Ligation to a new adaptor

1. Ligate 2 μl (200 ng) of the cut difference product (*Protocol 14A*, step 3) to N-adaptors in a 0.5 ml microcentrifuge tube by adding 3 μl of 10 × ligase buffer, 2 μl of 1 mg ml^{-1} N-12 oligo, 2 μl of 2 mg ml^{-1} N-24 oligo and 20 μl of water.
2. Anneal the oligos in a PCR machine by heating to 50 °C for 1 min, then cooling to 10 °C at 1 °C min^{-1}.
3. Add 1 μl of T4 DNA ligase, and incubate the reaction overnight at 14 °C.
4. Dilute the ligation to 1.25 ng μl^{-1} by addition of 130 μl of TE.

As the hybridization is assembled, take great care to remember which is the correct driver for the DP1 tester. The N-24 primer is used for amplification. The DP2 should look appreciably different from DP1 when run on an ethidium-stained gel. Under most circumstances, DP2 should produce clear bands within a background smear that will vary between experiments. These bands are usually real differences (representing either absolutely different or upregulated mRNAs), and can be cut out of the gel and cloned. If, however, a significant background smear remains it is possible either to repeat the DP2 at a higher stringency, or to proceed to a third difference product.

Protocol 15

Generation of further difference products (DP2 and DP3)

Equipment and reagents

- See *Protocols 12* and *13*

A Generating a second difference product (DP2)

1. Mix 10 μl (12.5 ng) of N-ligated DP1 (from *Protocol 14*) with 20 μl (10 μg) of driver (a driver/tester ratio of 800:1), and proceed through the subtraction hybridization procedure in Protocol 12.
2. Perform the initial amplification and MBN digestion by following Protocol 13A and 14B, using N-24 primers in place of J-24, and performing all annealing and extensions during the PCR at 72 °C.
3. Generate the final DP2 following Protocol 13C. Perform two reactions for each sample. Combine the products of the two reactions at step 6, and double the volumes of reagents used for extraction and precipitation.
4. Resuspend the pellet from the combined final PCR reactions at 0.5 μg μl^{-1} (usually 50 μl) in TE. This is the DP2.

B Generating a third difference product (DP3)[a]

1. Digest and ligate the DP2 to a J adaptor following Protocol 14, using J-adaptors instead of N-adaptors.
2. Dilute the J-ligated DP2 to 1 ng μl^{-1}. Set up hybridizations following Protocol 12, using driver to tester ratios between 5000:1 and 40 000:1. Prepare these hybridizations by keeping the driver concentration constant at 10 μg per hybridization, while varying the quantity of J-ligated tester.
3. Generate a DP3 following *Protocols 13* and *15A*, performing the final amplification for 23 cycles.
4. Resuspend the pellet from the two combined final PCR reactions at 0.5 μg μl^{-1} in TE. This is the DP3.

[a] It is unusual to have to proceed to a DP3. The ratio of driver/tester in DP2 can be varied from 400:1 to 2400:1. This may be preferable to generating a third difference product where products may be lost.

5 Common problems with cDNA RDA

The identification of many small problems and their possible solutions are dealt with in the footnotes of the protocol; however, most general problems relate to either low yield or to high background.

5.1 Low yield

Low yields of representations or difference products are typically due to including insufficient template in the PCR reactions. Insufficient template in the representation can result from excessively degraded mRNA or poor quality cDNA. Misjudgement of DNA concentrations, failed preliminary reactions (digestion, ligation, and hybridization) or loss of material during the procedure can all account for the lack of amplifiable template in the PCRs, which generate either the representation or the difference products. A representation or DP1 showing no yield at all is usually due to an error in setting up the PCR, or from loss of the DNA pellet (particularly during the steps leading up to digestion with MBN). The reactions should then be repeated using template from the postligation or post-MBN stages. When a low yield is encountered in the DP1 (or DP2), and the representation was at the correct concentration, the efficiency of the digestion and ligation should be checked. This is most easily accomplished by running uncut, cut, and religated representation (or DP) on an ethidium bromide stained 1.5% agarose gel and noting how the prominent bands shift down by 48 bp after digestion and back up again after ligation. The use of poor quality components or the presence of excessive salt in the DNA can cause failure of these reactions. Another potential cause of low yield is failure to resuspend the DNA properly before or after the subtractive hybridization.

5.2 High background

The presence of an excessively high background smear after two rounds of subtraction may be due to failure of the MBN digestion, or to incomplete melting of the DNA during the hybridization. In this case, the temperature of the PCR block should be accurately determined. In experiments derived from very small amounts of material, it is also possible that a high background indicates large-scale differences in quality between the driver and tester used for the subtraction.

The cloning of excessive numbers of false positives usually indicates that too low a stringency has been applied, while a lack of any bands in DP2 after obtaining potential products in DP1 probably indicates that too high a stringency has been used. The use of positive controls is advisable for first attempts at cDNA RDA. These can be performed by spiking the original tester mRNA or cDNA with a titrated quantity of an amplifiable *DpnII* gene fragment known to be absent from the driver. The addition of too many positive control species, however, is counter-productive as it may interfere with amplification of desired unknown species. The isolation of large numbers of clones that fail to produce matches with sequences in the DNA databases can indicate contamination of the starting cDNA with genomic DNA.

6 Screening of difference products

6.1 Primary screening

The products of the cDNA RDA procedure are known as difference products and consist of exponentially enriched PCR products derived from digested fragments of the cDNA that gave rise to the tester representation. Primary screening of difference products typically involves three steps: verification of a successful subtraction, identification of the products, and establishment of reproducibility.

A successful subtraction will generate over 90% genuine difference products; however, it is common for a visible band on an agarose gel to contain more than one true difference product. Also, where a low ratio of driver to tester has been employed, it is not unusual for a low background of products originating from highly abundant messages to persist. It is therefore necessary to clone the second difference products (DP2) as direct sequencing is seldom successful. Where the DP2 contains a mixture of fragments, the individual isolation of each different-sized band maximizes product recovery. After removal of the PCR adaptors by *Dpn*II digestion and gel purification, difference products can be ligated into the polylinker of a suitable plasmid vector that has been digested with a restriction enzyme (such as *Bam*HI or *Bgl*II) to generate compatible ends. The presence of inserts of the predicted size can be determined by PCR, using vector-specific primers. A rapid and convenient way to verify a cloned difference product is then to remove and label the insert for use as a probe on Southern blots of the original representations. This process also has the benefit of conserving precious RNA.

The advent of widely accessible and cheap automated sequencing, coupled with advanced search programs and ever expanding DNA sequence databases, provides the most straightforward method for the identification of cloned products. Cloned cDNA RDA products may be derived from any part of the cDNA, but are frequently located in coding regions and are not limited to the extreme 3′ end of the gene as are the products of differential display methods. Miniprepped plasmid, or, conveniently, the spin column-purified PCR products that result from insert confirmation, can therefore be sequenced directly. The sequences can then be compared with a DNA database, for example by 'Blast search' of GenBank using the algorithms developed by Altschul *et al.* (19). It is most productive to search in order the nucleotide database (blastn), the redundant protein database (blastx) and finally the expressed sequence tags (EST) database (tblastx). EST matches can be used as 'hooks' to walk along a gene until the open reading frame is discovered or the gene is identified. Typically, matches for the majority of products will be found in one of the databases.

On occasions, technical problems may arise which bias representations and prevent them reflecting accurately the mRNA from which they were generated.

Commonest of these problems is unequal RT in the generation of cDNA. Where a tester is derived from longer average length cDNA than a driver, restriction fragments occurring in the 5′-end of a transcript may be artificially enriched in the ensuing subtraction. These products will reflect a true difference between the representations, but not between starting mRNA populations. It is therefore necessary to generate separate preparations of RNA from identical cell populations so that the expression levels of difference products can be independently verified by Northern blot, or by RT–PCR using primers designed from the sequences of the difference products. Another way to test the reproducibility of difference product levels is to screen driver and tester specific cDNA libraries with products and observing their representations in each.

Protocol 16

Primary screening of difference products

Equipment and reagents

- *Dpn*II (NEB, 10 U μl^{-1}); *Bam*H1 (NEB 10 U μl^{-1}); calf intestinal alkaline phosphatase (CIP; 1 U μl^{-1}); T4 DNA ligase (NEB, 400 U μl^{-1})
- QIAquick DNA purification kit (Qiagen)
- pBluescriptKS^{+} vector (Stratagene)
- M13 forward and reverse primers

Method

1 Digest 12.5 μg of DP2 with *Dpn*II in a total volume of 300 μl. Extract the digest with PCI, then CI, precipitate with ethanol and resuspend the pellet in 100 μl of TE.

2 Prepare a 1.5% gel with wide lanes and load all 100 μl of cut DP2. Run the gel at a low voltage for several hours to separate the difference products.

3 Cut out each band with a clean scalpel blade and extract the DNA using a QIAquick DNA purification kit (*Protocol 11A*) or other suitable method.

4 Digest a suitable vector (e.g. pBluescriptKS^{+}) with *Bam*H1. Phosphatase the ends of the vector with CIP, and gel purify it (see step 3).

5 Ligate each difference product into the *Bam*H1 cut vector in a total volume of 5 μl using 50 ng of vector, 10–50 ng of insert, 0.5 μl of 10 × ligation buffer, and 0.5 μl of T4 DNA ligase.

6 Transform 3 μl of the ligation into DH5α and grow the bacteria on agar plates containing ampicillin.

7 Prepare a PCR reaction tube for 10 colonies per transformation.

8 Pick the colonies, touch them on to a gridded amp plate to regrow, and place the remainder of the colony into the PCR reactions.

9 Amplify the inserts using the M13 forward and reverse primers.

10 Gel-purify products which contain inserts of the predicted size and sequence them directly.

Protocol 16 continued

11 Miniprep any interesting products.

12 Generate several representation blots by running 0.75 μg of all the required representations on a 1.5% agarose gel and Southern blotting them on to a nylon membrane.

13 Prepare ^{32}P-labelled probes from the inserts of the difference product plasmids. Hybridize the products to the representation blots to verify true difference products.

14 Screen all true difference products against fresh isolates of RNA from the sample populations, either by Northern blotting, or by RT–PCR (*Protocol 17*).

Protocol 17

Screening difference products by RT–PCR

Equipment and reagents

- PCR machine
- Total RNA (from *Protocol 1*)
- 5 × First Strand Buffer (e.g. SuperScript II buffer, Life Technologies)
- 5 mM dNTP mixture (5 mM each of dATP, dCTP, dGTP, TTP), random hexamer primer (100 pg μl^{-1}), RNase inhibitor (Promega, 20–40 U μl^{-1})
- RNAse-free water
- Reverse Transcriptase (e.g. SuperScript II, Life Technologies, 200 U μl^{-1})
- *Taq* DNA polymerase (e.g. Gibco BRL, 5 U μl^{-1}); 10 × PCR buffer (Gibco BRL); 4 mM dNTPs (4 mM each dATP, dCTP, dGTP and TTP); 50 mM $MgCl_2$
- Oligonucleotide primers to tubulin (tubU, tubL), difference product specific primers

A Reverse transcription

1 Isolate total RNA from a fresh sample following *Protocol 1*.

2 Estimate the concentration of RNA by spectrophotometer reading at 260 nm.[a]

3 Remove contaminating genomic DNA following *Protocol 2*, resuspending the pellet at 1 μg μl^{-1}.

4 Place 7.5 μl of DNase-treated RNA in a 0.5 ml microcentrifuge tube to form the '+RT' sample. Place 2.5 μl of RNA in another tube and add 5 μl of RNase-free water to form the '−RT' control.[b]

5 Heat the tubes to 50 °C for 5 min to disrupt secondary structure.

6 Add 4 μl of 5 × first strand buffer, 4 μl of 5 mM dNTPs, 2 μl of 0.1 M DTT, 0.5 μl of RNase inhibitor, 1 μl of random hexamers to each tube.

7 Add 1 μl of RT to each '+RT' sample. Add 1 μl of water to the '−RT' controls.

8 Incubate at room temperature for 8 min, then 41 °C for 1 h 30 min.

9 Denature the enzyme and melt the DNA at 95 °C for 6 min. Use directly for PCR, or store at −20 °C.

Protocol 17 continued

B PCR

1. Prepare a 1:10 dilution of the cDNA in water.
2. Establish the quantity of cDNA in each sample by performing a test PCR using tubulin primers. Set up 20 μl PCR reactions in 0.5 ml microcentrifuge tubes consisting of 13 μl of water, 2 μl of 10 × PCR buffer, 0.6 μl of 4 mM dNTPs, 0.8 μl of $MgCl_2$, 1 μl of tubU primer, 1 μl of tubL primer and 1 μl of template cDNA.
3. Overlay the reactions with a drop of mineral oil, and place the reactions in a PCR machine. Heat to 95 °C for 1 min, then 80 °C for 10 s. Pause the cycle and add 1 μl of *Taq* DNA polymerase while the tubes are at 80 °C.
4. Cycle the reactions at 95 °C for 1 min, 59 °C for 1 min, and 72 °C for 2 min, with a final extension of 10 min. Perform separate reactions for 18, 20, 22, 24, and 26 cycles to determine the optimal cycle number at which to judge the concentration of cDNA.[c]
5. Load 10 μl of the reaction on to a 1.2% agarose gel containing ethidium bromide. Run the gel until it is possible to estimate accurately the concentration of the products.
6. Design specific primers to the difference products. Follow steps 2–5, adjusting the quantity of template cDNA so that all samples are equal with respect to tubulin expression, and changing the annealing temperature depending on the sequences of the primers.

[a] If there is insufficient RNA to take a spectrophotometer reading, judge the RNA on pellet size.

[b] To conserve RNA.

[c] It is essential to judge the quantities in the linear range of amplification for each set of primers.

6.2 Secondary screening

Potential uses for difference products continue after verification, identification, and confirmation of differences. In secondary screening procedures, difference products can be used as probes for identifying full-length cDNA, either by screening cDNA libraries or as the starting fragment for 5′ and 3′-RACE. Where directionality can be deduced, products may be cloned into expression vectors and may function as anti-sense constructs. They can also be used to express protein fragments, or as the sequence basis for the synthesis of peptides for use in antibody generation, although care should be taken to check the products for PCR errors, which are not uncommon after the many cycles necessary for cDNA RDA.

7 Modifications to cDNA RDA

7.1 Competition

It is often known in advance that one or more products are expressed in the tester but not in the driver. Known differences can provide a useful positive control for cDNA RDA, but they can also compete with other, more interesting, differences and reduce their amplification. Fortunately, addition of these known gene fragments into the driver will prevent, or at least significantly reduce, their appearance in the final DP2, increasing the chances of identifying new genes. Driver supplementation can be accomplished either by adding plasmids or PCR products containing full or partial cDNA for the undesired genes, or by including only the amplifiable portions of those genes. The quantity of competitor necessary for successful competition should be determined experimentally, but up to 50% of the mass of the driver can be composed of specific competitor sequences, with only a twofold drop in overall stringency. Because the specific competitor/target ratio far exceeds that of the driver/tester, amplification of the undesired target is effectively prevented. We have found that even highly abundant species can be successfully quenched by this approach. It is sometimes sufficient to include competitor in the second round of subtraction only, but care should be taken to also maintain a high driver/tester ratio to drive out other abundant cDNAs.

Some subtractions may be expected to contain a large number of differences. On these occasions, the difference products identified in DP2 are not exhaustive because some gene fragments are concealed by more vigorously amplifying fragments. In such cases it is possible to take advantage of competition to perform iterative cDNA RDA, in which differences (and false products) cloned from the initial DP2 are combined, ligated to R-adaptors, and amplified. This competitor product can then be mixed with the existing driver to drive out previously identified products and permit the amplification of a new set of true differences.

Protocol 18

Generation of a competitor amplicon

Equipment and reagents

- See *Protocols 12* and *13*
- Primers specific for known genes

Method

1. Identify genes that are known to be absent from the driver (DPs, or known genes).
2. Amplify the relevant portions of the genes by PCR using sequence-specific primers.
3. Gel purify the products and digest them with *Dpn*II.
4. Ligate 1 μg of the digested fragments to R-adaptors (*Protocol 8B*).
5. Amplify the *Dpn*II fragments following *Protocol 9A*.
6. Digest the R-adaptors with *Dpn*II (*Protocol 10*).
7. Resuspend the competitor amplicon at 0.5 $\mu g\ \mu l^{-1}$ in TE. Add the competitor into the driver during the subtractive hybridization (*Protocol 12*), keeping the amount of total driver plus competitor constant at 10 μg.[a]

[a] Competitor can form up to 50% of the driver with only a twofold drop in the driver/tester ratio.

7.2 Combination of representations

The ability to combine representations during cDNA RDA is a considerable advantage, as it allows subtractions to be targeted to identify specific subsets of potentially interesting differences. Representations can be combined to form the driver when the aim is to obtain genes that are expressed in one population, but not in several others. Conversely, testers can be prepared from combinations of representations in order to detect, for example, all the genes activated during a time-course experiment. Experimental strategies can become fairly complex when multiple drivers, testers, and competitors are in use, and a handy way to describe and present them is in the form of a Venn diagram. In combinatorial experiments, the driver/tester ratios of individual fragments will vary depending on the composition of the driver, and experimentation with ratios may be necessary to achieve the desired outcome.

7.2.1 Combined driver

Virtually any cDNA can be included in the driver without detrimental effects on the procedure, providing the average length is at least equal to that of the tester cDNA. As an example, it may be desirable to isolate genes that are transcribed in a wild-type cell line, but not in a mutant. To accurately target the subtraction to candidate genes normally under the control of the mutant gene, a mutant line complemented with the wild-type gene should be used as tester rather than the wild type itself. To ensure that differences are likely to be true targets of the gene in question, it is advisable, where possible, to prepare representations from other lines mutant for the same gene and to combine these representations to form the driver. Also, it is likely that the complemented line will contain a few known differences (possibly including the mutant gene itself or a selectable marker gene), and a competitor amplicon should be prepared from these and added to the driver. The subsequent subtraction will therefore only identify previously unknown differences that are expressed in the complemented line and absent from a range of mutant lines. These products would then be good candidates for screening against a number of different complemented mutant lines to establish reproducibility.

7.2.2 Combined tester

Combination of tester representations can be particularly useful for experiments which require the study of multiple samples, including post 'event' or developmental time courses or multiple treatments of the same cells or tissues (e.g. dose responses). The driver in these types of studies is normally the zero time or untreated sample, while the tester can be combined from representations generated from multiple time points or treatments. Depending on the tester composition, genes can be identified on the basis of early or late expression, or high and low doses. Genes expressed throughout the course of the experiment will be detected if all the treated representations are combined to form the tester. The difference products can then be used as probes on Southern blots of the individual representations that comprise the tester to give an indication of the peak of expression for each product.

7.3 Variable stringency

The term stringency used in the context of cDNA RDA subtractions refers to the driver/tester ratio, rather than to hybridization conditions of salt concentration and temperature, which remain constant. Standard, 'full stringency' driver/tester ratios for cDNA RDA are 100:1 for the first subtraction, and 800:1 for the second. When a cDNA RDA subtraction is performed at full stringency, differentially expressed gene fragments are typically detected with a false positive rate of about 10%. It is often found that not all the difference products are absolute, but that some are upregulated in the tester relative to the driver. If the stringency is decreased intentionally (or by accidental misjudgement of representation concentrations), more relative differences can be detected. Low stringency cDNA RDA has the capacity to detect genes that show as little as a three- to fivefold upregulation in the tester. However, only a subset of relative differences is likely to be detected, and if a comprehensive population is desired, other differential screening approaches may be more appropriate. cDNA RDA performed at too low a stringency is likely to result in a significant increase in the rate of false positives, evidenced by a smeary background on an ethidium bromide-stained gel at DP2. One way to overcome the cloning of false products is to experiment with the driver/tester ratio until the desired compromise between relative differences and false positives is achieved. This can be accomplished either by repeating the generation of DP2 at ratios between 400:1 and 2400:1 or by generating a third difference product (DP3). The prescribed stringency for the DP3 hybridization is 40 000:1, conditions which generally favour the production of absolute differences with a very low false positive rate. At this very high stringency, however, many potentially interesting sequences can be lost, and a compromise driver/tester ratio between 5000:1 and 10 000:1 may produce better results.

7.4 Increasing cDNA coverage

If a subtraction fails to generate any amplifiable difference products after DP2, despite experimental alterations of stringency, and positive controls have been successful, it is likely that there are very few, if any, major differences between the populations. Although the proportion of cDNA molecules covered by representations prepared from *Dpn*II digested cDNA is already very high (about 86%), it is possible to repeat the experiment using a representation generated from cDNA cut with a restriction enzyme which recognizes a different 4-bp recognition site, such as *Nla*III. It is preferable to select enzymes that leave 4-bp 5′ overhangs, as these anneal and ligate more efficiently, and the terminal sequence of the 12-mer oligonucleotides should be redesigned to be compatible with the new site. An alternative approach is to generate first strand cDNA using an oligo(dT) primer that incorporates a *Dpn*II site. This will then permit representation of cDNA molecules that contain only one *Dpn*II site, provided that this site occurs between 130 bp and 1.2 kb from the 3′ end of the gene. An oligo(dT) primer containing a *Dpn*II site should also be anchored to the sequence preceding the poly(A) tail by inclusion of at least two bases of random sequence at its 5′end. This ensures that all fragments bounded at their 3′ ends by the primer *Dpn*II site are of equal length, and are consequently likely to behave in a similar manner during hybridizations.

7.5 cDNA RDA in prokaryotes

Prokaryotes pose special difficulties because they produce predominantly short, unstable, non-polyadenylated mRNA. To overcome these problems, cDNA is prepared by random-priming from total RNA, and an additional step is taken to deplete representations of ribosomal RNA. Codon usage in prokaryotes is significantly different to eukaryotic usage, and *Dpn*II may not be the optimal enzyme to use when preparing representations. Because of the shorter average length and instability of prokaryotic messages, several different four-cutter enzymes may be necessary to generate several representations before sufficient coverage is attained. Subtractions must then be performed separately with each of these representations. The abbreviated protocol provided is subject to modifications due to its very recent development.

Protocol 19
cDNA RDA with prokaryotes

Equipment and reagents

- See *Protocols 5–15*
- Primers specific for known bacterial rRNA genes
- Random hexamer primers (100 pg μl^{-1})
- Genomic-Tip 100/G system (Qiagen)
- R-, J-, and N-adaptors, with the final 4 bp of the 12-mers modified to complement the 4-bp overhang of the chosen restriction enzyme
- Suitable 4 base pair-cutting restriction enzymes

Method

1. Isolate bacterial chromosomal DNA, for example using the Genomic-Tip 100/G system.
2. Use synthetic oligonucleotide primers, based on the rRNA sequences of the bacteria being studied, to amplify the rRNA genes from the DNA prepared in step 1.
3. Digest the rRNA gene products with *Dpn*II or another suitable four-cutting restriction enzyme, ligate on R-linkers and generate a competitor amplicon (see *Protocol 18*).
4. Prepare bacterial total RNA by a reliable method.[a]
5. Generate first strand cDNA using random hexamer primers instead of oligo(dT) (*Protocols 5* and *6*).

Protocol 19 continued

5 Perform second strand synthesis as in *Protocols* 5 or 6.

6 Generate bacterial cDNA representations and perform cDNA RDA following *Protocols* 7–15, using 50% competitor rRNA amplicon in all of the subtractive hybridizations.[b]

[a] For example, Bowler *et al.* (16).

[b] It may be necessary to perform four rounds of subtraction. If so, perform the final round (to DP4) at a driver/tester ratio of 200 000:1.

References

1. Sagerström, C. G., Sun, B. I., and Sive, H. L. (1997). *Annu. Rev. Biochem.* **66**, 751.
2. Lisitsyn, N. A., Lisitsyn, N. M., and Wigler, M. (1993). *Science* **259**, 946.
3. Lisitsyn, N. A. and Wigler, M. (1995). In *Methods in enzymology* (ed. Vogt, P. K. and Verma, I. M.)Vol. **254**, p. 291. Academic Press, London.
4. Hubank, M. and Schatz, D. G. (1994). *Nucleic Acids Res.* **22**, 5640.
5. Welford, S. M., Gregg, J., Chen, E., Garrison, D., Sorensen, P. H., Denny, C. T., and Nelson, S. F. (1998). *Nucleic Acids Res.*, **26**, 3059.
6. Geng, M., Wallrapp, C., MullerPillasch, F., Frohme, M., Hoheisel, J. D., and Gress, T. M. (1998). *Biotechniques*, **25**, 434.
7. Lewis, B. C., Shim, H., Li, Q., Wu, C. S., Lee, L. A., Maity, A., and Dang, C. V. (1997). *Molec. Cell. Biol.*, **17**, 4967.
8. Wang, X. Z., Kuroda, M., Sok, J., Batchvarova, N., Kimmel, R., Chung, P., Zinszner, H., and Ron, D. (1998). *EMBO J.*, **17**, 3619.
9. Zhu, Y. X., Zhang, Y. F., Luo, J. C., Davies, P. J., and Ho, D. T.H. (1998). *Gene*, **208**, 1.
10. Melià, M-J., Bofill, N., Hubank, M., and Meseguer, A. (1998). *Endocrinology*, **139**, 688.
11. Zheng, W. P. and Flavell, R. A. (1997). *Cell*, **89**, 587.
12. Lawson, N. D. and Berliner, N. (1998). *Proc. Natl. Acad. Sci. USA*, **95**, 10129.
13. Kellam, P. (1998). *Trends In Microbiology*, **6**, 160.
14. Reiter, R. E., Gu, Z. N., Watabe, T., Thomas, G., Szigeti, K., Davis, E., Wahl, M., Nisitani, S., Yamashiro, J., LeBeau, M. M., Loda, M., and Witte, O. N. (1998). *Proc. Natl. Acad. Sci. USA*, **95**, 1735.
15. Lucas, S., DeSmet, C., Arden, K. C., Viars, C. S., Lethe, B., Lurquin, C., and Boon, T. (1998).*Cancer Research*, **58**, 743.
16. Bowler, L. D., Hubank, M., and Spratt, B. G. (1999) *Microbiology, UK*, **145**, 3529–3537.
17. Chomczynski, P. and Sacchi, N. (1987). *Anal. Biochem.*, **162**, 156.
18. Sambrook, I., Fritsch, E. F., and Maniatis, T. (ed.) (1989) *Molecular Cloning: a laboratory approach* (2nd edn), Chapter 6, p. 6.20. Cold Spring Harbor Laboratory Press, New York.
19. Altschul, S. F., Madden, T. L., Schaffer, A. A., Zhang, J., Zhang, Z., Miller , W., and Lipman, D. J. (1997) *Nucleic Acids Res.* **25**, 3389.
20. Hubank, M. and Schatz, D. G. (1999) In *Methods in Enzymology* (ed. Weissman S. M.) **303**, 325–349. Academic Press, London.

Chapter 5

Identification of differentially expressed genes by suppression subtractive hybridization

Sejal Desai, Jason Hill, Stephanie Trelogan,
Luda Diatchenko and Paul D. Siebert
CLONTECH Laboratories Inc., Palo Alto, CA, USA

1 Introduction

Suppression subtractive hybridization (SSH) is a polymerase chain reaction (PCR)-based cDNA subtraction method. This powerful technique can be used to compare two mRNA populations and obtain cDNAs of genes that are either over-expressed or exclusively expressed in one population compared with another.

SSH features several improvements over other cDNA subtraction methods that have been described in the literature (1–5). SSH eliminates any intermediate steps for physical separation of single-stranded (ss) and double-stranded (ds) cDNAs, requires only one round of subtractive hybridization, and can achieve greater than a 1000-fold enrichment for differentially expressed cDNAs (6–8). To overcome the problem of differences in mRNA abundance, SSH incorporates a hybridization step that normalizes (equalizes) sequence abundance by standard hybridization kinetics during the course of subtraction. This dramatically increases the probability of obtaining low-abundance differentially expressed cDNAs and simplifies analysis of the subtracted library. The SSH technique has many potential applications in molecular genetics and positional cloning studies, including the identification of disease-related, developmental, tissue-specific, and other differentially expressed genes.

In this chapter, we provide complete protocols for generating subtracted cDNA libraries and instructions for differential screening of these libraries. As an example, we have included a comprehensive characterization of four subtracted, tumour-specific libraries. These four subtracted libraries, as well as several previously reported subtracted libraries, serve as the basis for a discussion of the advantages and limitations of SSH.

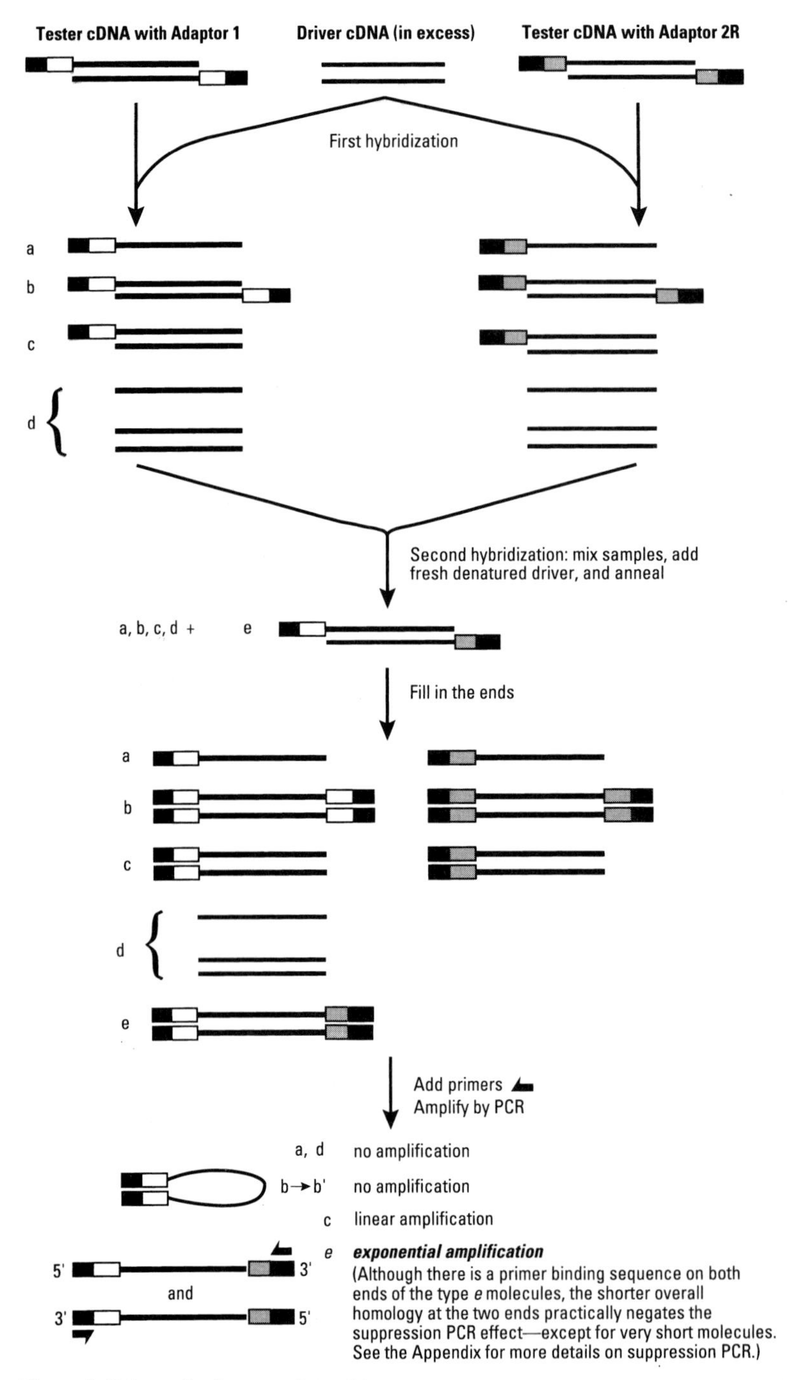

Figure 1 Schematic diagram of the SSH procedure. Solid boxes represent the outer part of the adaptor Ad1 and Ad2R, and correspond to the PCR primer 1 (P1) sequence. Clear boxes represent the inner part of adaptor Ad1 and correspond to nested PCR primers 1 (NP1). Shaded boxes represent the inner part of adaptor Ad2R and correspond to nested PCR primer 2R (NP2R).

2 The principle of SSH

The SSH procedure includes several steps. First, cDNA is synthesized from the two types of tissues or cell populations being compared. The cDNA population in which specific transcripts are to be found is called tester cDNA, and the reference cDNA population is called driver cDNA. For cDNA synthesis, the conventional method described by Gubler and Hoffman (9) can be used. In the second step, tester and driver cDNAs are digested with a four-base-cutting restriction enzyme that yields blunt ends. The tester cDNA is then subdivided into two portions, and each is ligated to a different ds adaptor (adaptors 1 (Ad1) and 2R (Ad2R)). The ends of the adaptors are not phosphorylated, so only one strand of each adaptor is covalently attached to the 5′ ends of the cDNAs.

The molecular events of subtractive hybridization and selective amplification of differentially expressed genes are illustrated in *Figure 1*. In the first hybridization, an excess of driver cDNA is added to each sample of tester cDNA. The samples are then heat-denatured and allowed to anneal. *Figure 1* shows the type A, B, C, and D molecules generated in each sample. Type A molecules, which represent ss tester molecules, include equal concentrations of high- and low-abundance sequences because reannealing is faster for more abundant molecules due to the second-order kinetics of hybridization (10). At the same time, type A molecules are significantly enriched for differentially expressed sequences because common non-target cDNAs form type C molecules with the driver. During the second hybridization, the two primary hybridization samples are mixed together in the presence of fresh denatured driver. Type A cDNAs from each tester sample are now able to associate and form type B, type C, and new type E hybrids. Type E hybrids are double-stranded tester molecules with different ss ends that correspond to Ad1 and Ad2R. Freshly denatured driver cDNA is added to enrich fraction E further for differentially expressed sequences.

The entire population of molecules is then subjected to two rounds of PCR to amplify selectively the differentially expressed sequences. Prior to the first cycle of primary PCR, the adaptor ends are filled in, creating the complementary primer binding sites needed for amplification. Type A and D molecules lack primer annealing sites and cannot be amplified. Type B molecules form a panhandle-like structure that suppresses amplification (11, 12). Type C molecules have only one primer annealing site and can only be amplified linearly. Only type E molecules, which have two different primer annealing sites, can be amplified exponentially. These differentially expressed sequences are greatly enriched in the final subtracted cDNA pool.

3 Protocols

3.1 Materials and reagents

3.1.1 Oligonucleotides

The following oligonucleotides are used at a concentration of 10 μM. Whenever possible, oligonucleotides should be gel-purified.

- cDNA synthesis primers: 5′-TTTTGTACAAGCTT$_{30}$-3′
- Ad1: 5′-CTAATACGACTCACTATAGGGCTCGAGCGGCCGCCCGGGCAGGT-3′; 3′-GGCCCGTCCA-5′
- Ad2R: 5′-CTAATACGACTCACTATAGGGCAGCGTGGTCGCGGCCGAGGT-3′; 3′-GCCGGCTCCA-5′
- Primer 1 (P1): 5′-CTAATACGACTCACTATAGGGC-3′
- Nested primer 1 (NP1): 5′-TCGAGCGGCCGCCCGGGCAGGT-3′
- Nested primer 2R (NP2R): 5′-AGCGTGGTCGCGGCCGAGGT-3′
- Glyceraldehyde 3-phosphate dehydrogenase 5′ primer: 5′-ACCACAGTCCATG-CCATCAC-3′
- G3PDH 3′ primer: 5′-TCCACCACCCTGTTGCTGTA-3′
- Blocking solution: A mixture of the cDNA synthesis primer, nested primers (NP1 and NP2R), and their respective complementary oligonucleotides (2 mg ml^{-1} each).

3.1.2 Buffers and enzymes

All chemical reagents were obtained from Sigma.

(1) First-strand synthesis: AMV reverse transcriptase (20 units μl^{-1}; Life Technologies), 5 × first-strand buffer (250 mM Tris–HCl pH 8.5, 40 mM $MgCl_2$, 150 mM KCl, and 5 mM dithiothreitol).

(2) Second-strand synthesis: 20 × second-strand enzyme cocktail (DNA polymerase I, 6 units μl^{-1}, New England Biolabs; RNase H, 0.25 units μl^{-1}, Epicentre Technologies; *Escherichia coli* DNA ligase, 1.2 units μl^{-1}, New England Biolabs); 5 × second-strand buffer (500 mM KCl, 50 mM ammonium sulphate, 25 mM $MgCl_2$, 0.75 mM β-Nicotinamide adenine dinucleotide (β-NAD), 100 mM Tris–HCl pH 7.5, 0.25 mg ml^{-1} BSA); and T4 DNA polymerase (3 units μl^{-1}, New England Biolabs).

(3) Endonuclease digestion: 10 × *Rsa*I restriction buffer (100 mM Bis Tris Propane–HCl pH 7.0, 100 mM $MgCl_2$ and 1 mM dithiothreitol (DTT)), *Rsa*I (10 units μl^{-1}, New England Biolabs).

(4) Adaptor ligation: T4 DNA ligase (400 units μl^{-1}: contains 3 mM ATP, New England Biolabs), 5 × DNA ligation buffer (250 mM Tris–HCl pH 7.8, 50 mM $MgCl_2$, 10 mM DTT, 0.25 mg ml^{-1} BSA).

(5) Hybridization: 4 × hybridization buffer (4 M NaCl, 200 mM HEPES pH 8.3, 4 mM cetyltrimethyl ammonium bromide (CTAB)); dilution buffer (20 mM HEPES–HCl pH 8.3, 50 mM NaCl, 0.2 mM ethylenediamine tetraacetic acid (EDTA)).

(6) Polymerase chain reaction (PCR) amplification: Advantage, cDNA PCR Mix (CLONTECH Laboratories, Inc.). This mix contains a mixture of KlenTaq-1 and DeepVent, DNA polymerases (New England Bio Labs) and TaqStart™ Antibody (CLONTECH); 10 × reaction buffer (40 mM Tricine–KOH (pH 9.2 at 22 °C), 3.5 mM $Mg(OAc)_2$, 10 mM KOAc, 75 mg ml^{-1} bovine serum albumin (BSA)). TaqStart Antibody provides automatic hot start PCR (13). Alternatively, *Taq* DNA polymerase alone can be used, but five additional thermal cycles will be

needed in both the primary and secondary PCR, and the additional cycles may cause higher background. If the Advantage cDNA PCR Mix is not used, manual hot start or hot start with wax beads is strongly recommended to reduce non-specific DNA synthesis.

(7) General reagents: dNTP mix (10 mM each dNTP, Amersham Pharmacia Biotech), 20 × EDTA/glycogen mix (0.2 M EDTA; 1 mg ml^{-1} glycogen), 4 M NH_4OAc, sterile H_2O, and ExpressHyb™ Hybridization Solution (CLONTECH)

Please note that all cycling parameters are optimized for a Perkin-Elmer DNA Thermal Cycler 480 (Perkin-Elmer). Perkin-Elmer GeneAmp® PCR Systems 2400/9600 have also been tested. For the latter, the denaturing time needs to be reduced from 30 s to 10 s. For a different type of thermal cycler, the cycling parameters must be optimized for that machine.

3.2 Preparation of the subtracted cDNA library

3.2.1 cDNA synthesis

There are two steps for cDNA synthesis: first-strand cDNA synthesis and second-strand cDNA synthesis. During the first step, AMV reverse transcriptase is used to synthesize cDNA with poly(A)$^+$ RNA as a template. During the second step, DNA polymerase I is used to synthesize second-strand cDNA with first-strand cDNA as a template. The following protocol is recommended for generating a subtracted library from 2 μg of poly(A)$^+$ RNA. If enough poly(A)$^+$ RNA is not available, the switching mechanism at the 5′ end of RNA templates (SMART™) amplification technology (CLONTECH) can be used to preamplify high-quality cDNA from total RNA (14).

Protocol 1

First-strand cDNA synthesis

Equipment and reagents

- First-strand buffer: 250 mM Tris–HCl pH 8.5, 40 mM $MgCl_2$, 150 mM KCl, 5 mM dithiothreitol
- cDNA synthesis primer
- Sterile water
- dNTP mixture (10 mM each)
- 0.1 M DTT
- AMV reverse transcriptase (20 units μl^{-1})

Method

Perform this procedure individually with each tester and driver poly(A)$^+$ RNA sample.

1. For each tester and driver sample, combine the following components in a sterile 0.5-ml microcentrifuge tube (do not use a polystyrene tube): poly(A)$^+$ RNA (2 μg) to 2–4 μl; cDNA synthesis primer (10 μM) to 1 μl. If needed, add sterile H_2O to a final volume of 5 μl.
2. Incubate the tubes at 70 °C in a thermal cycler for 2 min.

Protocol 1 continued

3 Cool at room temperature for 2 min and briefly centrifuge using a PicoFuge, microcentrifuge (Stratagene) at maximum rotation speed (6400 r.p.m.).

4 Add the following to each reaction tube: 5 × first-strand buffer (2 μl); dNTP mixture (10 mM each) (1 μl); sterile H_2O^a (0.5 μl); 0.1 M DTT (0.5 μl); AMV reverse transcriptase (20 units μl^{-1}) (1 μl).

5 Gently vortex and briefly centrifuge the tubes.

6 Incubate the tubes at 42 °C for 1.5 h in an air incubator.

7 Place the tubes on ice to terminate first-strand cDNA synthesis and immediately proceed to second-strand cDNA synthesis.

[a] Optional: To monitor the progress of cDNA synthesis, dilute 0.5 μl of [α-^{32}P]-dCTP (10 mCi ml^{-1}, 3000 Ci $mmol^{-1}$) with 9 μl of H_2O and replace the H_2O above with 1 μl of the diluted label.

Protocol 2

Second-strand cDNA synthesis

Equipment and reagents

- Second-strand buffer: 500 mM KCl, 50 mM ammonium sulphate, 25 mM $MgCl_2$, 0.75 mM β-NAD, 100 mM Tris–HCl pH 7.5, 0.25 mg ml^{-1} BSA
- Second-strand enzyme cocktail: DNA polymerase I, 6 units μl^{-1}; RNase H, 0.25 units μl^{-1}; *Escherichia coli* DNA ligase, 1.2 units μl^{-1}
- dNTP mix (10 mM)
- T4 DNA polymerase
- 0.2 M EDTA
- 4 M NH_4OAc

Method

1 Add the following components (previously cooled on ice) to the first-strand synthesis reaction tubes: sterile H_2O (48.4 μl); 5 × second-strand buffer (16.0 μl); dNTP mix (10 mM) 1.6 μl); 20 × second-strand enzyme cocktail (4.0 μl).

2 Mix the contents and briefly spin the tubes. The final volume should be 80 μl.

3 Incubate the tubes at 16 °C (water bath or thermal cycler) for 2 h.

4 Add 2 μl (6 units) of T4 DNA polymerase. Mix contents well.

5 Incubate the tube at 16 °C for 30 min in a water bath or a thermal cycler.

6 Add 4 μl of 0.2 M EDTA to terminate second-strand synthesis.

7 Perform phenol–chloroform extraction and ethanol precipitation. We recommend that you use 0.5 volume of 4 M NH_4OAc rather than NaOAc for ethanol precipitation.

8 Dissolve pellet in 50 μl of H_2O.

9 Transfer 6 μl to a fresh microcentrifuge tube. Store this sample at –20 °C until after *Rsa*I digestion for agarose gel electrophoresis to estimate yield and size range of the ds cDNA products synthesized.

3.2.2 *Rsa*I digestion

Perform the following procedure with each experimental ds tester and driver cDNA. This step generates shorter, blunt-ended ds cDNA fragments optimal for subtractive hybridization.

Protocol 3

*Rsa*I digestion

Equipment and reagents

- ds cDNA (43.5 μl)
- *Rsa*I (10 units μl^{-1})
- 0.2 M EDTA
- *Rsa*I restriction buffer: 100 mM Bis Tris Propane-HCl pH 7.0, 100 mM $MgCl_2$, and 1 mM DTT

Method

1. Add the following reagents into the tube from *Protocol 2*, step 8 above: ds cDNA (43.5 μl); 10 × *Rsa*I restriction buffer (5.0 μl); *Rsa*I (10 units μl^{-1}) (1.5 μl).
2. Mix and incubate at 37 °C for 2 h.
3. Use 5 μl of the digest mixture and analyse on a 2% agarose gel along with undigested cDNA (from Section 3.2.1, *Protocol 2*, step 9) to analyse the efficiency of *Rsa*I digestion.

Note: continue the digestion during electrophoresis and terminate the reaction only after you are satisfied with the results of the analysis.

4. Add 2.5 μl of 0.2 M EDTA to terminate the reaction.
5. Perform phenol–chloroform extraction and ethanol precipitation.
6. Dissolve the pellet in 5.5 μl of H_2O and store at −20 °C.

3.2.3 Adaptor ligation

It is strongly recommended that you perform subtractions in both directions for each tester/driver cDNA pair. Forward subtraction is designed to enrich for differentially expressed transcripts present in tester but not in driver; reverse subtraction is designed to enrich for differentially expressed sequences present in driver but not in tester. The availability of such forward- and reverse-subtracted cDNAs will be useful for differential screening of the resulting subtracted tester cDNA library (see Section 3.3).

The tester cDNAs are ligated separately to Ad1 (tester 1–1 and 2–1) and Ad2R (tester 1–2 and 2–2). It is highly recommended that a third ligation of both adaptors 1 and 2R to the tester cDNAs (unsubtracted tester control) be performed and used as a negative control for subtraction.

Please note that the adaptors are not ligated to the driver cDNA.

Protocol 4

Adaptor ligation

Equipment and reagents

- DNA ligation buffer: 250 mM Tris–HCl pH 7.8, 50 mM $MgCl_2$, 10 mM DTT, 0.25 mg ml^{-1} BSA
- Sterile water
- 3 mM ATP
- T4 DNA ligase
- 0.2 M EDTA

Method

1 Dilute 1 μl of each *Rsa*I-digested tester cDNA from the above section with 5 μl of sterile H_2O.

2 Prepare a master ligation mix of the following components for each reaction: sterile H_2O (2 μl); 5 × ligation buffer (2 μl); adenosine triphosphate (ATP) (3 mM) (1 μl); T4 DNA ligase (400 units μl^{-1}) (1 μl).

4 For each tester cDNA mixture, combine the following reagents in a 0.5-ml microcentrifuge tube in the order shown. Pipette the solution up and down to mix thoroughly.

Tube no.	**1**	**2**
Component	**Tester 1–1 (μl)**	**Tester 1–2 (μl)**
Diluted tester cDNA	2	2
Adaptor Ad1 (10 μM)	2	–
Adaptor Ad2R (10 μM)	–	2
Master ligation mix	6	6
Final volume	10	10

5 In a fresh microcentrifuge tube, mix 2 μl of tester 1–1 and 2 μl of tester 1–2. This is your unsubtracted tester control. Do the same for each tester cDNA sample. After ligation, approximately one-third of the cDNA molecules in each unsubtracted tester control tube will have two different adaptors on their ends, suitable for exponential PCR amplification with adaptor-derived primers.

6 Centrifuge the tubes briefly and incubate at 16 °C overnight.

7 Stop the ligation reaction by adding 1 μl of 0.2 M EDTA.

8 Heat samples at 72 °C for 5 min to inactivate the ligase.

9 Briefly centrifuge the tubes. Remove 1 μl from each unsubtracted tester control and dilute into 1 ml of H_2O. These samples will be used for PCR amplification (Section 3.2.6).

Preparation of your experimental Adaptor-Ligated Tester cDNAs 1–1 and 1–2 is now complete.

3.2.4 Ligation efficiency test

The following PCR experiment is recommended to verify that at least 25% of the cDNAs have adaptors on both ends. This experiment is designed to amplify fragments that span the adaptor/cDNA junctions of testers 1–1 and 1–2 by adaptor-specific P1 primer and a gene-specific primer. Theoretically, the PCR products generated using one gene-specific primer and the adaptor-specific primer should be about the same intensity on an agarose/EtBr gel as the PCR products amplified using two gene-specific primers. It is important that the amplified gene-specific fragment does not contain an *Rsa*I restriction site. The selected G3PDH primers, listed in Section 3.1.1, work well for human, mouse, and rat cDNA samples.

Protocol 5

Analysis of ligation

Equipment and reagents

- Tester 1–1 (ligated to Ad1)
- Tester 1–2 (ligated to Ad2R)
- G3PDH 3′ primer (10 μM)
- G3PDH 5′ primer (10 μM)
- PCR primer P1 (10 μM)
- Master mix: sterile water, 10 × PCR reaction buffer, dNTP mix (10 mM), 50 × Advantage cDNA PCR Mix
- Mineral oil
- 2% agarose/EtBr gel

Method

1 Dilute 1 μl of each ligated cDNA from Section 3.2.3 (step 8) (e.g. the testers 1–1 and 1–2) into 200 μl of H_2O.

2 Combine the reagents in four separate tubes as follows:

	Tube			
Component (in μl)	**1**	**2**	**3**	**4**
Tester 1–1 (ligated to Ad1)	1	1	–	–
Tester 1–2 (ligated to Ad2R)	–	–	1	1
G3PDH 3′ primer (10 μM)	1	1	1	1
G3PDH 5′ primer (10 μM)	–	1	–	1
PCR primer P1 (10 μM)[a]	1	–	1	–
Total volume	3	3	3	3

3 Prepare a master mix for all of the reaction tubes plus one additional tube. For each reaction, combine the reagents in the following order:

Reagent	**Amount per reaction tube (μl)**	**Amount for 4 reactions (μl)**
Sterile H_2O	18.5	92.5
10 × PCR reaction buffer	2.5	12.5
dNTP mix (10 mM)	0.5	2.5
50 × Advantage cDNA PCR Mix	0.5	2.5
Total volume	22.0	110.0

4 Mix thoroughly and briefly centrifuge the tubes.

5 Aliquot 22 μl of master mix into each reaction tube from step 2.

Protocol 5 continued

6 Overlay with 50 μl of mineral oil. Skip this step if an oil-free thermal cycler is used.

7 Incubate the reaction mixture in a thermal cycler at 75 °C for 5 min to extend the adaptors. (Do not remove the samples from the thermal cycler.)

8 Immediately commence 20 cycles of: 94 °C, 30 s; 65 °C, 30 s; 68 °C, 2.5 min.

9 Examine the products by electrophoresis on a 2% agarose/EtBr gel.

[a] Primer P1 contains 22 nucleotides corresponding to the 5′ end sequence of both adaptors Ad1 and Ad2R.

Typical results are shown in *Figure 2*. If no products are visible after 20 cycles, perform five more cycles of amplification, and again analyse the product by agarose gel electrophoresis. The number of cycles will depend on the abundance of the specific gene. The efficiency of ligation is determined by the relative intensity of the bands corresponding to the PCR products of tube 2 to 1 for adaptor Ad1 and 4 to 3 for adaptor Ad2R. A fourfold or more difference in intensity reflects a ligation efficiency of 25% or less and will substantially reduce the subsequent subtraction efficiency. In this case, the ligation reaction should be repeated with fresh samples before proceeding to the next step.

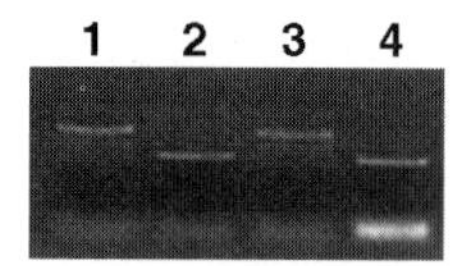

Figure 2 Results of the ligation efficiency analysis. Lanes 1–4: ligation efficiency analysis for human liver-metastasized colon tumour sample. Lane 1: PCR products using Tester 1-1 (Adaptor 1-ligated liver-metastasized colon carcinoma cDNA tester) as the template and the G3PDH 3′ primer and PCR primer 1. Lane 2: PCR products using Tester 1-1 (Adaptor 1-ligated liver-metastasized colon carinoma cDNA tester) as the template, and the G3PDH 3′ and 5′ primers. Lane 3: PCR products using Tester 1-2 (Adaptor 2R-ligated liver-metastasized colon carcinoma cDNA tester) as the template, and the G3PDH 3′ primer and PCR primer 1. Lane 4: PCR products using Tester 1-2 (Adaptor 2R-ligated liver-metastasized colon carcinoma cDNA tester) as the template, and the G3PDH 3′ and 5′ primers. Shown is an EtBr-stained 2% agarose gel.

For mouse or rat cDNAs, the PCR products amplified with the G3PDH 3′ primer and PCR primer P1 will be ~1.2 kb instead of the 0.75-kb band observed for human cDNA (because rat and mouse G3PDH cDNAs lack the *Rsa*I restriction site in the 340 nt position). However, for the human cDNA (which contains the *Rsa*I site), the presence of a 1.2-kb band suggests that the cDNAs are not completely digested by *Rsa*I. If a significant amount of this longer PCR product persists, the procedure should be repeated from the step of *Rsa*I digestion (Section 3.2.2).

3.2.5 Subtractive hybridization

Protocol 6

First hybridization

Equipment and reagents

- *Rsa*I-digested driver cDNA (*Protocol* 3, step 6)
- Ad1-ligated Tester 1-1 (*Protocol* 4, step 8)
- Ad2R-ligated Tester 1-2 (*Protocol* 4, step 8)
- Hybridization buffer: 4 M NaCl, 200 mM HEPES pH 8.3, 4 mM CTAB
- Mineral oil

Method

1 For each tester sample, combine the reagents in the following order:

Component	Hybridization sample 1 (μl)	Hybridization sample 2 (μl)
*Rsa*I-digested driver cDNA (*Protocol* 3, step 7)	1.5	1.5
Ad1-ligated Tester 1-1 (*Protocol* 4, step 8)	1.5	–
Ad2R-ligated Tester 1-2 (*Protocol* 4, step 8)	–	1.5
4 × hybridization buffer	1.0	1.0
Final volume	4.0	4.0

2 Overlay samples with one drop of mineral oil and centrifuge briefly.

3 Incubate samples in a thermal cycler at 98 °C for 1.5 min.

4 Incubate samples at 68 °C for 7–12 h and then proceed immediately to the second hybridization.

Note: The protocol uses 15 ng of ligated tester cDNA and 450 ng of driver cDNA. The ratio of driver to tester can be changed if a different subtraction efficiency is desired (see Section 5.6).

Protocol 7

Second hybridization

Equipment and reagents

- Driver cDNA
- Hybridization buffer: 4 M NaCl, 200 mM HEPES pH 8.3, 4 mM CTAB
- Mineral oil
- Dilution buffer: 20 mM HEPES–HCl pH 8.3, 50 mM NaCl, 0.2 mM EDTA

Method

Repeat the following steps for each experimental driver cDNA

1 Add the following reagents into a sterile 0.5-μl microcentrifuge tube: driver cDNA (*Protocol* 3, step 6), 1 μl; 4 × hybridization buffer, 0.3 μl.

2 Place 1 μl of this mixture in a 0.5-ml microcentrifuge tube and overlay it with one drop of mineral oil.

Protocol 7 continued

3 Incubate in a thermal cycler at 98 °C for 1.5 min.
4 Remove the tube of freshly denatured driver from the thermal cycler.
5 To the tube of freshly denatured driver cDNA, add hybridized sample 1.1 and hybridized sample 1.2 (from the first hybridization in *Protocol* 6) in that order. This ensures that the two hybridization samples are mixed only in the presence of excess driver cDNA.
6 Incubate the hybridization reaction at 68 °C overnight.
7 Add 200 μl of dilution buffer to the tube and mix well by pipetting.
8 Incubate in a thermal cycler at 68 °C for 7 min.

3.2.6 PCR amplification: Selection of differentially expressed cDNAs

Each experiment should have at least four reactions: (i) subtracted tester cDNAs; (ii) unsubtracted tester control; (iii) reverse-subtracted tester cDNAs; and (iv) unsubtracted driver control for the reverse subtraction.

Protocol 8

Primary PCR

Equipment and reagents

- Master mix: sterile H_2O, 10 × PCR reaction buffer, dNTP mix (10 mM), PCR primer P1 (10 μM), 50 × Advantage cDNA PCR Mix
- Mineral oil
- 2% agarose/EtBr gel
- TAE buffer

Method

1 Aliquot 1 μl of each diluted cDNA sample (i.e. each subtracted sample from Section 3.2.5, *Protocol* 7, step 7, and the corresponding diluted unsubtracted tester control from Section 3.2.3, *Protocol* 4, step 9) into an appropriately labelled tube.
2 Prepare a master mix for all of the primary PCR tubes plus one additional tube. For each reaction combine the reagents in the order shown:

Reagent	**Amount per reaction (μl)**
Sterile H_2O	19.5
10 × PCR reaction buffer	2.5
dNTP mix (10 mM)	0.5
PCR primer P1 (10 μM)	1.0
50 × Advantage cDNA PCR Mix	0.5
Total volume	24.0

3 Aliquot 24 μl of master mix into each reaction tube prepared in step 1 above.
4 Overlay with 50 μl of mineral oil. Skip this step if an oil-free thermal cycler is used.

Protocol 8 continued

5 Incubate the reaction mixture in a thermal cycler at 75 °C for 5 min to extend the adaptors. Do not remove the samples from the thermal cycler.

6 Immediately commence 27 cycles of: 94 °C, 30 s; 66 °C, 30 s; 72 °C, 1.5 min.

7 Analyse 8 μl from each tube on a 2.0% agarose/EtBr gel run in 1 × (0.4 m Tris acetate, pH 8.1 and 0.01 m EDTA, Na_2) buffer.

Please note:

- The sequence for PCR primer P1 is present at the 5′ ends of both adaptors Ad1 and Ad2R and, therefore, can be used in a single primer PCR amplification
- If no PCR product is observed after 27 cycles, amplify for three more cycles, and analyse the products again by gel electrophoresis
- If no PCR product is observed in the subtracted or unsubtracted (unsubtracted tester control) samples, the activity of the *Taq* polymerase needs to be examined. If the problem is not with the polymerase mix, optimize the PCR cycling parameters by decreasing the annealing and extension temperatures in small increments. Lowering the temperature by only 1 °C can dramatically increase the background. Initially, try reducing the annealing temperature from 66 °C to 65 °C and the extension temperature from 72 °C to 71 °C
- If PCR products are observed in the unsubtracted (unsubtracted tester control) samples but not in the subtracted sample, proceed to *Protocol 9* and perform more cycles of secondary PCR

Protocol 9

Secondary PCR

Equipment and reagents

- Master mix: sterile H_2O, 10 × PCR reaction buffer, nested PCR primer NP1 (10 μM), nested PCR primer NP2R (10 μM), dNTP mix (10 mM), 50 × Advantage cDNA PCR Mix
- Primary PCR mixture
- Mineral oil
- 2% agarose/EtBr gel

Method

1 Dilute 3 μl of each primary PCR mixture in 27 μl of H_2O.

2 Aliquot 1 μl of each diluted primary PCR product mixture from step 1 into an appropriately labelled tube.

3 Prepare a master mix for the secondary PCR samples plus one additional reaction by combining the reagents in the following order:

Reagent	Amount per reaction (μl)
Sterile H_2O	18.5
10 × PCR reaction buffer	2.5
Nested PCR primer NP1 (10 μM)	1.0
Nested PCR primer NP2R (10 μM)	1.0
dNTP mix (10 mM)	0.5
50 × Advantage cDNA PCR Mix	0.5
Total volume	24.0

Protocol 9 continued

4 Aliquot 24 μl of master mix into each reaction tube from step 2.

5 Overlay with one drop of mineral oil. Skip this step if an oil-free thermal cycler is used.

6 Immediately commence 10–12 cycles of: 94 °C, 30 s; 68 °C, 30 s; 72 °C, 1.5 min.

7 Analyse 8 μl from each reaction on a 2.0% agarose/EtBr gel.

Please note: The secondary PCR products of subtracted samples usually look like smears with or without a number of distinct bands (*Figure 3*). If no product is observed after 12 cycles, perform three more cycles of amplification, and again check the products by gel electrophoresis. Add cycles sparingly—too many cycles will increase background.

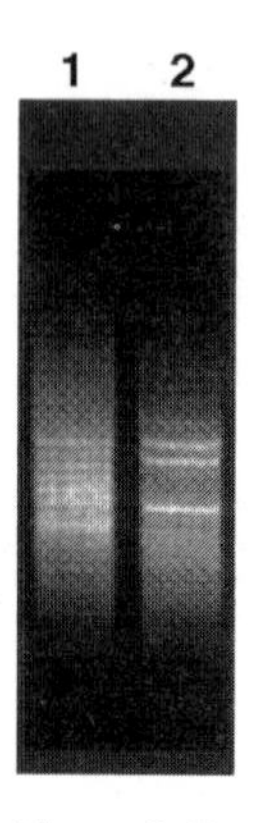

Figure 3 Results of secondary PCR. Secondary PCR was performed as described in the protocol using unsubtracted (lane 1) and subtracted (lane 2) diluted primary PCR from the liver-metastasized colon carcinoma sample subtracted against normal liver. The cDNA smear ranges from 0.2 kb to 1.5 kb.

3.2.7 Subtraction efficiency test

At this point, it is important to determine the efficiency of the SSH procedure by comparing the abundance of known cDNAs before and after subtraction. Ideally, you should use a gene that is not differentially expressed (e.g. a housekeeping gene) and a gene previously demonstrated to be differentially expressed between the two RNA sources. These comparisons can be performed using either PCR or hybridization techniques.

The test described below uses the G3PDH primers to confirm the reduced relative abundance of G3PDH following the SSH procedure.

Figure 4 shows an example of G3PDH reduction in a lung tumour-specific library. In general, for the unsubtracted cDNA, a G3PDH product is observed after 18–23 cycles, depending on the abundance of G3PDH in the particular tissue or cells. For example, in skeletal muscle and heart poly(A)$^+$ RNA, G3PDH is extremely abundant. However, in the subtracted samples, a product should be observed about five to 15 cycles later than in the unsubtracted samples.

Protocol 10

PCR analysis of subtraction efficiency

Equipment and reagents

- Diluted subtracted cDNA (2nd PCR product)
- Diluted unsubtracted control (2nd PCR product)
- G3PDH 5′ primer (10 μM)
- G3PDH 3′ primer (10 μM)
- Sterile H_2O
- 10 × PCR reaction buffer
- dNTP mix (10 mM)
- 50 × Advantage cDNA PCR Mix
- Mineral oil
- 2% agarose/EtBr gel

Method

1 Dilute the subtracted and unsubtracted (unsubtracted tester control 1-c and 2-c) secondary PCR products 10-fold in H_2O.

2 Combine the following reagents in 0.5-ml microcentrifuge tubes in the order shown:

Tube no.	1 (μl)	2 (μl)
Diluted subtracted cDNA (2nd PCR product)	1.0	–
Diluted unsubtracted control (2nd PCR product)	–	1.0
G3PDH 5′ primer (10 μM)	1.2	1.2
G3PDH 3′ primer (10 μM)	1.2	1.2
Sterile H_2O	22.4	22.4
10 × PCR reaction buffer	3.0	3.0
dNTP mix (10 mM)	0.6	0.6
50 × Advantage cDNA PCR Mix	0.6	0.6
Total volume	30.0	30.0

3 Mix and briefly centrifuge the tubes.

4 Overlay with one drop of mineral oil. Skip this step if an oil-free thermal cycler is used.

5 Use the following thermal cycling program for 18 cycles: 94 °C, 30 s; 60 °C, 30 s; 68 °C, 2 min.

6 Remove 5 μl from each reaction, place it in a clean tube, and store on ice. Put the rest of the reaction back into the thermal cycler for three more cycles.

7 Repeat step 6 three times (i.e. remove 5 μl after 24, 27, and 30 cycles).

8 Examine the 5-μl samples (i.e. the aliquots that were removed from each reaction after 18, 21, 24, 27, and 30 cycles) on a 2.0% agarose/EtBr gel (*Figure 4*).

As a positive control for the enrichment of differentially expressed genes, repeat the PCR procedure above using PCR primers for a gene known to be expressed in the tester RNA, but not in the driver RNA. This cDNA should be

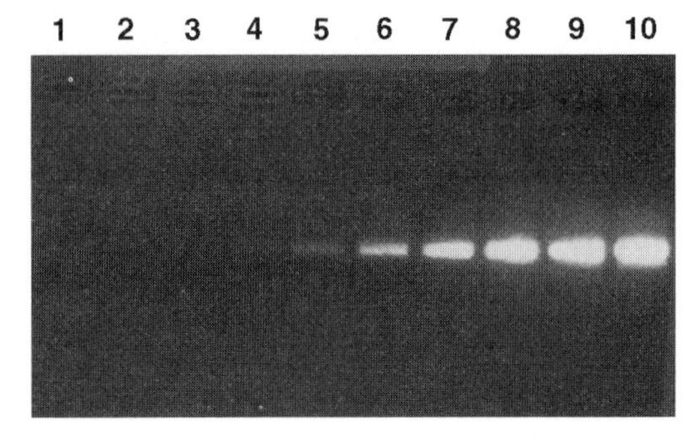

Figure 4 Reduction of G3PDH abundance by SSH. PCR was performed on lung tumour subtracted (lanes 1–5) or unsubtracted (lanes 6–10) secondary PCR products with the G3PDH 5′ and 3′ primers. Lanes 1 and 6: 18 cycles; lanes 2 and 7: 21 cycles; lanes 3 and 8: 24 cycles; lanes 4 and 9: 27 cycles; and lanes 5 and 10: 30 cycles.

enriched by the subtraction procedure. Do not use PCR primers that amplify a cDNA fragment that contains an *Rsa*I restriction site between the PCR priming sites because it will not be amplified due to *Rsa*I digestion prior to the subtraction procedure.

3.2.8 Cloning of subtracted cDNAs

Once a subtracted sample has been confirmed to be enriched in cDNAs derived from differentially expressed genes, the PCR products (from Section 3.2.6, secondary PCR) can be subcloned using several conventional cloning techniques. The following describes two such methods that are currently used.

T/A cloning

Use 3 μl of the secondary PCR product (from *Protocol 9*, step 6) for cloning with a T/A-based system, such as the AdvanTAge® PCR Cloning Kit (CLONTECH), according to the manufacturer's protocol. The library is transformed into bacteria (electrocompetent cells) by electroporation (1.8 kV) using a pulser (BioRad) and plated on to agar plates containing ampicillin, X-Gal, and IPTG. Recombinant (white colonies) clones are picked and used to inoculate LB medium in 96-well microtitre plates. Bacteria should be allowed to grow at 37 °C for 4 h before insert amplification (*Protocol 11*). Typically, 10^4 independent clones from 1 μl of secondary PCR product can be obtained using the above cloning system and electroporation. It is important to optimize the cloning efficiency because a low cloning efficiency will result in a high background.

Site-specific or blunt-end cloning

For site-specific cloning, cleave at the *Eag*I, *Not*I, and *Xma* (*Sma*I, *Srf*I) sites embedded in the adaptor sequences and then ligate the products into an appropriate plasmid vector. Keep in mind that all of these sites might be present in the cDNA fragments. The *Rsa*I site in the adaptor sequences can also be used for blunt-ended cloning. Commercially available cloning kits are suitable for these purposes.

The number of independent clones obtained for each library depends on the estimated number of differentially expressed genes, as well as the subtraction and subcloning efficiencies. In general, 500 colonies can be initially arrayed and studied. The complexity of the library can be increased by additional subcloning of the secondary PCR products (from Section 3.2.6).

3.3 Differential screening of the subtracted cDNA library

Two approaches can be utilized for differential screening of the arrayed subtracted cDNA clones: cDNA dot blots and colony hybridization. For colony hybridization, bacterial colonies are transferred to nylon filters and lysed using conventional protocols. This method is usually cheaper and more convenient, but it is less sensitive and gives a higher background than PCR-based cDNA dot blots. The cDNA array approach is highly recommended (*Protocol 11*).

3.3.1 Amplification of cDNA inserts by PCR

For high-throughput screening, a 96-well format PCR from one of several thermal cycler manufacturers is recommended. Alternatively, single tubes can be used.

Protocol 11

cDNA insert analysis by PCR

Equipment and reagents

- Master mix: 10 × PCR reaction buffer, nested primer NP1, nested primer NP2R, dNTP Mix (10 mM), H_2O, 50 × Advantage cDNA PCR Mix
- LB-amp medium
- 2% agarose/EtBr gel

Method

1. Randomly pick 96 white bacterial colonies.
2. Grow each colony in 100 μl of LB-amp medium in a 96-well plate at 37 °C for at least 2 h (up to overnight) with gentle shaking.
3. Prepare a master mix for 100 PCR reactions:

	Per reaction (μl)
10 × PCR reaction buffer	2.0
Nested Primer NP1[a]	0.6
Nested Primer NP2R[a]	0.6
dNTP Mix (10 mM)	0.4
H_2O	15.0
50 × Advantage cDNA PCR Mix	0.4
Total volume	19.0

4. Aliquot 19 μl of the master mix into each tube or well of the reaction plate.

Protocol 11 continued

5 Transfer 1 μl of each bacterial culture (from step 2, above) to each tube or well containing master mix.

6 Perform PCR in an oil-free thermal cycler with the following conditions:
   - 1 cycle: 94 °C, 2 min
   - then 22 cycles: 94 °C, 30 s; 68 °C, 3 min

7 Analyse 5 μl from each reaction on a 2.0% agarose/EtBr gel (*Figure* 5).

[a] Alternatively, primers flanking the insertion site of the vector can be used in PCR amplification of the inserts.

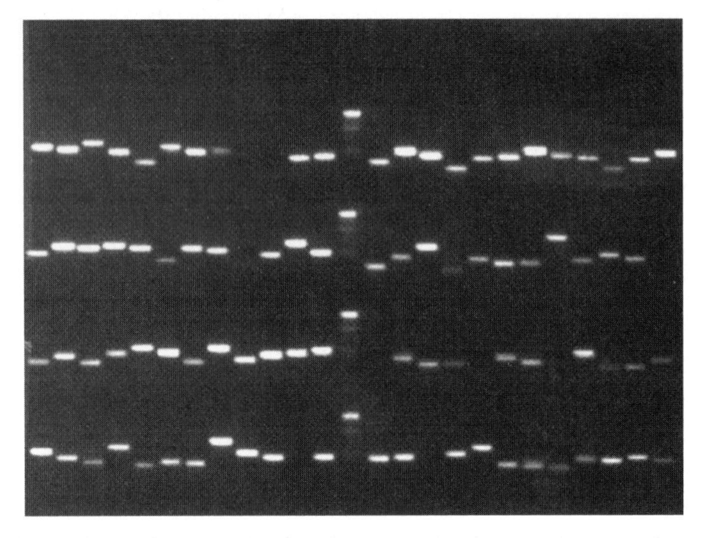

Figure 5 Insert screening analysis. Ninety-six randomly picked white colonies from lung tumour-specific subtracted clones were subjected to PCR (see *Protocol 11*) and analysed on a 2.0% agarose/EtBr gel.

3.3.2 Preparation of cDNA dot blots of the PCR products

Protocol 12

Preparation of cDNA dot blots

Equipment and reagents

- PCR product
- 0.6 N NaOH
- 0.5 M Tris–HCl (pH 7.5)
- 2 × SSC (300 mM NaCl and 30 mM Na_3 Citrate × $2H_2O$)
- UV cross-linking device (Stratagene)

Method

1 For each PCR reaction, combine 5 μl of the PCR product and 5 μl of 0.6 N NaOH (freshly made or at least freshly diluted from concentrated stock).

Protocol 12 continued

2 Transfer 1–2 μl of each mixture to a nylon membrane. This can be accomplished by dipping a 96-well replicator in the corresponding wells of a microtitre dish used in the PCR amplification and spotting it on to a dry nylon filter. Make at least two identical blots for hybridization with subtracted and reverse-subtracted probes (see Section 3.2.3).

Note: We highly recommend that you make four identical blots. Two of the blots will be hybridized to forward- and reverse- subtracted cDNAs and the other two can be hybridized to cDNA probes synthesized from tester and driver mRNAs.

3 Neutralize the blots for 2–4 min in 0.5 M Tris–HCl (pH 7.5) and wash in 2 × SSC.

4 Immobilize cDNA on the membrane using a UV crosslinking device (such as Stratagene's UV Stratalinker), or bake the blots under vacuum at 80 °C for 2–4 h.

3.3.3 Differential hybridization with tester and driver cDNA probes

Label tester and driver cDNA probes by random-primer labelling using a commercially available kit. The hybridization conditions given here have been optimized for CLONTECH's ExpressHyb solution; the optimal hybridization for other systems should be determined empirically.

The following four different probes will be used for differential screening hybridization.

(1) Tester-specific subtracted probe (forward-subtracted probe).
(2) Driver-specific subtracted probe (reverse-subtracted probe).
(3) cDNA probe synthesized directly from tester mRNA.
(4) cDNA probe synthesized directly from driver mRNA.

Protocol 13

Differential hybridization

Equipment and reagents

- Prehybridization solution: SSC, sheared salmon sperm DNA (10 mg ml^{-1}), blocking solution (containing 2 mg ml^{-1} of unpurified NP1, NP2R, cDNA synthesis primers and their complementary oligonucleotides)
- ExpressHyb Hybridization Solution (CLONTECH)
- Low-stringency (2 × SSC/0.5% SDS) washing buffer
- Hybridization probes: 20 × SSC, 50 μl of sheared salmon sperm DNA (10 mg ml^{-1}), 10 μl blocking solution, purified probe (at least 10^7 c.p.m. per 100 ng of subtracted cDNA)
- High-stringency (0.2 × SSC/0.5% SDS) washing buffer

Method

1 Prepare a prehybridization solution for each membrane:

Protocol 13 continued

(a) Combine 50 μl of 20 × SSC, 50 μl of sheared salmon sperm DNA (10 mg ml^{-1}), and 10 μl of blocking solution (containing 2 mg ml^{-1} of unpurified NP1, NP2R, cDNA synthesis primers and their complementary oligonucleotides).

(b) Boil the blocking solution for 5 min, then chill on ice.

(c) Combine the blocking solution with 5 ml of ExpressHyb Hybridization Solution.

2 Place each membrane in the prehybridization solution prepared in step 1. Prehybridize for 40–60 min with continuous agitation at 72 °C.

Note: It is important that you add blocking solution in prehybridization solution. As subtracted probes contain the same adaptor sequences as arrayed clones, these probes hybridize to all arrayed clones regardless of the sequences.

3 Prepare hybridization probes:

(a) Mix 50 μl of 20 × SSC, 50 μl of sheared salmon sperm DNA (10 mg ml^{-1}) and 10 μl blocking solution, and purified probe (at least 10^7 c.p.m. per 100 ng of subtracted cDNA). Make sure the specific activity of each probe is approximately equal.

(b) Boil the probe for 5 min, then chill on ice.

(c) Add the probe to the prehybridization solution.

4 Hybridize overnight with continuous agitation at 72 °C.

5 Prepare low-stringency (2 × SSC/0.5% SDS) and high-stringency (0.2 × SSC/0.5% SDS) washing buffers and warm them up to 68 °C.

6 Wash membranes with low-stringency buffer (4 × 20 min at 68 °C), then wash with high-stringency buffer (2 × 20 min at 68 °C).

7 Perform autoradiography.

8 If desired, remove probes from the membranes by boiling for 7 min in 0.5% SDS. Blots can typically be reused at least five times.

Note: To minimize hybridization background, store the membranes at −20 °C when they are not in use.

Note that the first two probes are the secondary PCR products (*Protocol* 9, step 6) of the subtracted cDNA pool. The last two cDNA probes can be synthesized from the tester and driver $poly(A)^+$ RNA. They can be used as either single-stranded or double-stranded cDNA probes (*Protocols 1* and *2*). Alternatively, unsubtracted tester and driver cDNA (*Protocol* 9, step 6) or preamplified cDNA from total RNA (14) can be used if not enough $poly(A)^+$ RNA is available.

4 Results

Tumour-specific genes are involved in the dysfunction of many normal processes, including cell proliferation (15), cell–cell communication (16, 17), cell–matrix interactions (18, 19), tumour invasion, metastasis (20, 21), and senescence (22,

23). Altered expression of such genes may facilitate initiation or progression of a neoplasm as oncogenes do, or may inhibit it as do tumour suppressor genes. Thus, identification of differentially expressed genes may provide a better understanding of the underlying molecular mechanisms of a particular biological process or disease state, as well as revealing the functions of the genes themselves.

In this study, we performed subtractive hybridization of several tumour tissues against their corresponding normal tissues, and vice versa. Our tumour samples included lung carcinoma, adenocarcinoma of prostate, adenocarcinoma of breast, and post-colorectal adenocarcinoma (liver-metastasized colon carcinoma). Subtracted cDNAs specific for tumour and normal tissues were obtained using the SSH protocol as described in previous sections of this chapter. Two hybridization conditions were attempted for each tumour-specific and normal tissue-specific subtraction: driver/tester ratios of 30:1 and 50:1 were performed side by side in the first hybridization step (see Section 3.2.5). For each subtraction, we chose one subtracted cDNA pool based on the efficiency of reduction of G3PDH (see *Protocol 10*); each cDNA pool was cloned in T/A cloning vector. We randomly picked 288 white colonies (three 96-well plates) from each library, except for the liver-metastasized colon carcinoma-specific library, from which we selected 480 white colonies (five 96-well plates). All clones were subjected to differential screening analysis (see protocols in Section 3.3); those clones that were judged to be differentially expressed were sequenced. Differential screening analysis was performed until all differentially expressed cDNAs on a given 96-well plate had already been identified on previous plates.

4.1 Subtraction efficiency analysis

We determined the efficiency of the SSH procedure by comparing the abundance of G3PDH in the subtracted and unsubtracted cDNA pools (see *Protocol 10*) for each experimental condition. Because G3PDH is a housekeeping gene, it should be subtracted out during the SSH procedure. For each tumour-specific and normal tissue-specific subtraction, one subtracted pool, obtained with either the 30:1 or 50:1 driver/tester ratio, was chosen based on the best reduction of G3PDH after subtraction. *Figure 4* shows the G3PDH analysis of the lung tumour-specific cDNA pool obtained using a 50:1 driver/tester ratio. In the subtracted sample, the G3PDH product was observed about 15 cycles later than in the unsubtracted sample. This corresponds to approximately a 1000-fold reduction of G3PDH. For all subtractions except the breast tumour-specific subtraction, we found that a 50:1 ratio was optimum. For the breast tumour-specific subtraction, a ratio of 30:1 gave better results.

4.2 Differential screening analysis

Figure 6 shows a differential screening analysis of lung tumour tissue (tester) subtracted against normal lung tissue (driver). Equal amounts of PCR products from 96 clones randomly picked from the lung tumour-specific library were arrayed on nylon membranes for quadruplicate screening. Four identical membranes

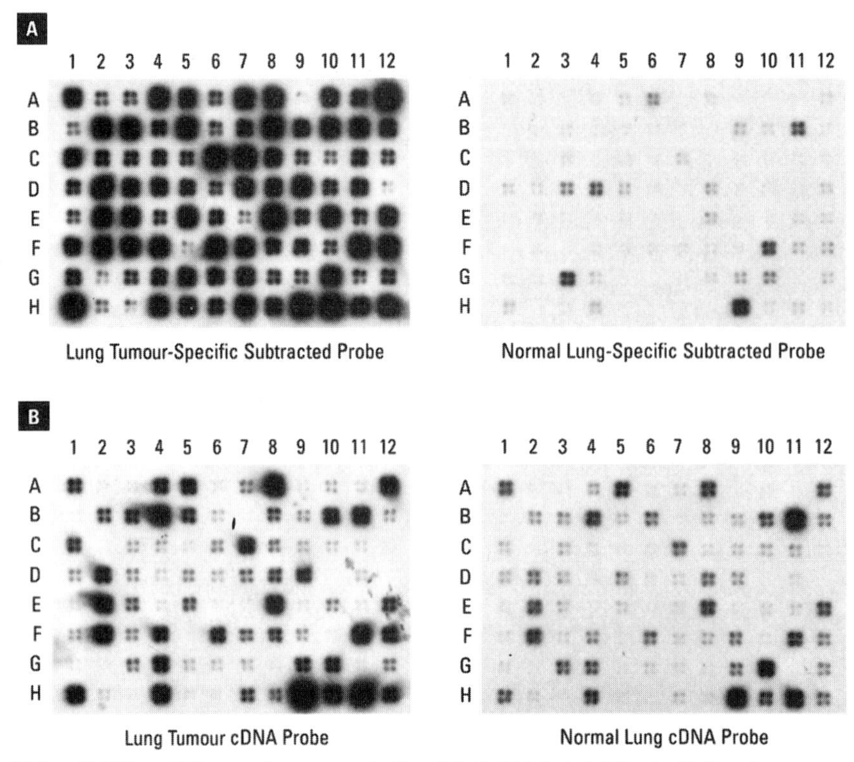

Figure 6 Differential screening approach. Panel A. Dot blots hybridized with lung tumour-specific subtracted probe (lung tumour = tester, normal lung = driver) and normal lung-specific subtracted probe (normal lung = tester, lung tumour = driver). Panel B. Dot blots hybridized with unsubtracted cDNA probes made from lung tumour and normal lung mRNA.

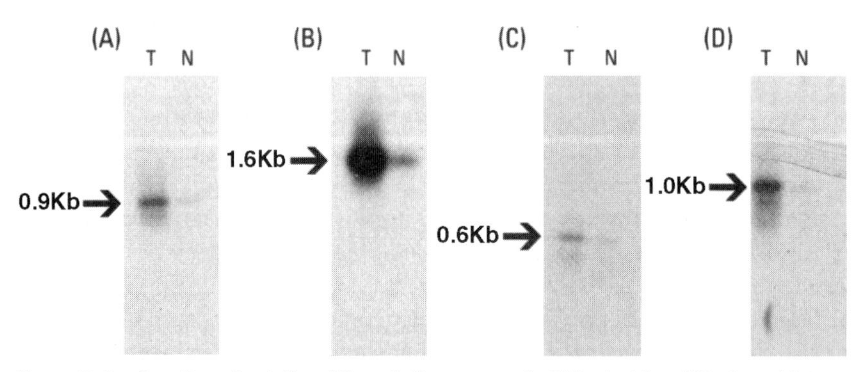

Figure 7 Confirmation of putative differentially expressed cDNAs by Virtual Northern blot analysis. Each lane contains 0.5 μg of SMART PCR-amplified cDNA from either lung tumour or normal lung samples (as indicated). To produce each blot, SMART cDNA was electrophoresed on a 1.0% agarose gel and transferred to a nylon membrane. EAch blot was hybridzed to one of the following ^{32}P-labelled cDNA clones: (A) anticarcinoma surface monoclonal antibody CR4E8 (IgM λ chain variable region), (B) human mRNA for IgV-L, (C) human retinally abundant regulator of G-protein signalling hRGS-r, and (D) unknown sequence. All were confirmed positive by Virtual Northern blot analysis. Exposure time: 1 hr to 2 days. G3PDH hybridization confirmed that lanes were loaded equally (not shown).

Table 1 Summary of differential screening analysis

Subtracted cDNA Libraries	% Differentially expressed clones Total	by unsubtracted probe	by subtracted probe	% of antibody clones in the subtracted library	examples of redundant genes in the subtracted library
Liver metastasized colon carcinoma	60	39	21	0	Non-specific crossreacting antigen (prevalence, 1/9), CD24 signal transducer and L1-cadherin (prevalence, 1/36)
Normal liver	50	45	5	0	Serum albumin (prevalence, 1/5), and haptoglobin (prevalence, 1/8)
Lung tumour	60	34	26	30	Fibronectin (prevalence, 1/72)
Normal lung	60	48	12	9	Surfactant proteins (prevalence, 1/6), and Glutathione peroxidase (prevalence, 1/32).
Prostate tumour	13	13	0	0	–
Normal prostate	3	3	0	0	–
Breast tumour	70	60	10	36	Fibronectin and Leupexin (prevalence, 1/36)

from each 96-well plate were prepared. Each of the four membranes was hybridized with four different probes. These include tester- and driver-specific subtracted probes, which are purified secondary PCR products of subtracted tester and driver (see Section 3.2.6, *Protocol 9*), and cDNA probes made directly from the original tester and driver mRNA (i.e. unsubtracted probes).

Table 1 shows a summary of the differential screening results from all the subtracted libraries. In the lung tumour-specific library, 60% of the clones were derived from upregulated differentially expressed genes. The remainder of the clones represent genes that are not differentially expressed. Using cDNA probes synthesized from tester and driver mRNA (*Figure 6B*), we found that 34% of the clones were differentially expressed; using subtracted probes (*Figure 6A*), we found that an additional 26% of the clones were differentially expressed.

The breast tumour-specific library contained the highest percentage of differentially expressed clones (70%), whereas prostate tumour- and normal prostate-specific libraries contained very few differentially expressed clones (13% and 3%, respectively).

4.3 Confirmation of differential expression

The differential expression of candidate clones identified by differential screening can be confirmed by Northern blot analysis, Virtual Northern blot analysis

(14, 24), or reverse transcription (RT)—PCR. We used Virtual Northern blot analysis as we had very little RNA available to work with. To create Virtual Northern blots, cDNAs that correspond to poly(A)$^+$ RNA were preamplified from 1.0 μg of total RNA (14). These preamplified cDNAs were then electrophoresed on 1.2% agarose gels and transferred to nylon membranes.

Several random clones identified by differential screening analysis as differentially expressed were chosen from each library and subjected to Virtual Northern blot analysis. Except for the prostate-specific libraries, our rate of confirmation of differential expression was very high. For the breast-, lung-, and liver-specific libraries, differential expression of 59 of 62 clones chosen using both subtracted and unsubtracted probes was confirmed by Virtual Northern blot analysis. Four of these clones from the lung tumour-specific library are shown in *Figure 7*. In prostate tumour- and normal prostate-specific libraries, none of the 24 clones identified by differential screening analysis using subtracted probes was confirmed by Virtual Northern blot analysis. Only those clones that were identified by differential screening analysis using unsubtracted cDNA probes synthesized from tester or driver mRNA were confirmed.

4.4 Sequence analysis of differentially expressed clones

Table 2 shows a summary of the number of independent differentially expressed clones isolated from each library, known and unknown sequences, and sequence homology. Of 257 non-redundant clones isolated from various subtracted libraries, 174 were known, 66 were unknown, and 17 had partial homology to known sequences. Most of the cDNA fragments were detected two to three times during differential screening. However, some of the cDNAs were significantly more prevalent in subtracted libraries than others (see *Table 1*).

Table 2 Sequence analysis of differentially expressed clones

	No. of non-redundant sequences					
Subtracted cDNA libraries	**known sequences[a]**	**partial homology to known sequences[b]**	**no significant homology[c]**	**total**	**No. of matches to genes of known function**	**No. of matches to genes associated with cancer**
Liver metastasized colon carcinoma	70	10	24	104	64	42
Liver normal	26	0	4	30	—	—
Lung tumour	23	2	16	41	25	16
Lung normal	26	2	6	34	27	5
Breast tumour	22	—	11	33	20	5
Prostate tumour	6	3	3	12	5	1
Prostate normal	1	—	2	3	1	1
Total	**174**	**17**	**66**	**257**	**132**	**70**

[a] > 95% homology to a known sequence.

[b] > 60% homology to a known sequence in 100 or more nucleotides.

[c] < 60% homology to a known sequence.

Many genes were found in more than one tumour-specific library. For example, mRNAs for ribosomal proteins were upregulated in liver-metastasized colon carcinoma-specific libraries and breast tumour-specific libraries. Proteosome subunits were also upregulated in lung tumour-, liver-metastasized colon carcinoma- and breast tumour-specific libraries. Finally, fibronectin mRNA was upregulated in the lung tumour-specific and breast tumour-specific libraries.

Of the 132 genes with known functions, 70 were known to be associated with cancer. For example, carcinoembryonic antigen (CEA) was found to be differentially expressed and highly prevalent in the liver-metastasized colon carcinoma-specific library. CEA expression is greatly increased in colon carcinomas, resulting in a rise in serum levels (25). Glutathione peroxidase, an antioxidant, was downregulated in lung tumour. This finding is consistent with other reports (26). Human prostate-specific antigen (PSA) gene, a known prostate tumour marker (27), was found to be downregulated in the prostate tumour-specific library.

Many antibody clones were found to be upregulated in lung and breast tumours compared with their normal tissue counterparts (see *Table 1*). It is unclear if this finding reflects only the infiltration of lung and breast tumours by B lymphocytes, or if these antibodies are specific for tumour-cell surface-specific antigens.

5 Discussion

Since its development in 1996 (6), the SSH technique has been applied in many different fields of research, in our laboratories (8, 24, 28, 29, and present study) and in others (30–34; also see www.clontech.com), for the construction of subtracted libraries. Here, we summarize the results of these studies, providing detailed information about the requirements for the starting material, complexity and specificity of the subtracted library. We also discuss the potential advantages and limitations of the SSH method.

5.1 Level of background and differential screening

Although the SSH method greatly enriches for differentially expressed genes, the subtracted sample will still contain some cDNAs that correspond to mRNAs common to both the tester and driver samples. The level of these non-differentially expressed cDNAs (i.e. background) in the subtracted pool depends somewhat on the quality of the starting RNA and the performance of the particular subtraction, but it mainly depends on the particular samples used as tester and driver in subtraction. A high level of background is usually observed when very few mRNA species are differentially expressed between tester and driver. In general, the smaller the number of differentially expressed mRNAs and the lower the quantitative difference in expression, the higher the background. That is why the percentage of differentially expressed clones in the subtracted library may vary significantly.

Several examples of this variability can be found in previous studies. In our testis-specific subtracted library, about 95% of the clones corresponded to testis-specific mRNAs (8). A subtracted library specific for transcripts expressed in the oestrogen-receptor positive MCF7 cell line, but not in the oestrogen-receptor negative MDA-MB-231 cell line, contained about 70% differentially expressed clones (30). A subtracted library specific for an activated Jurkat cell line consisted of 40% differentially expressed clones (7). Finally, transcripts associated with gamma-globin production, but not with beta-globin production, represented only 12% of the gamma-globin producing hybridoma-subtracted library (24). We believe that the low percentage of differentially expressed clones in the latter case reflects very small differences in transcription of the genes between the two stages of human–mouse hybridoma culture (24). A high level of background is also observed when the subtraction efficiency is low, which can be caused by the poor quality of starting material (see quality of the starting sample in this section), as in the case of the prostate (*Table 1*).

The above examples demonstrate that it is almost impossible to predict the subtraction efficiency of any particular tester and driver sample. We suggest that you analyze each subtracted library by differential screening, which allows you to determine the identity of cDNAs in the subtracted library that actually correspond to differentially expressed mRNAs. In differential screening, clones from the subtracted library are arrayed on a membrane and screened with four different probes (see *Protocol 12*): driver cDNA probe (unsubtracted), tester cDNA probe (unsubtracted), subtracted driver-specific probe, and subtracted tester-specific probe. The hybridization results of two unsubtracted and two subtracted probes are then compared. Tester-specific clones should hybridize to the tester probes but not to the driver probes, or should hybridize to the tester probes with higher efficiency. *Figure 6(A,B)* represents an example of differential screening analysis. Clones that are hybridized to the unsubtracted tester-specific probe (lung tumour cDNA probe), but not to the unsubtracted driver specific probe (normal lung cDNA probe), have a high probability (over 95%) of corresponding to differentially expressed mRNAs (e.g. *Figure 6*, clones A4 and A8). However, cDNA clones that correspond to a differentially expressed mRNA that is rare in the original tester may not detectably hybridize to the unsubtracted cDNA probe as the concentration of these cDNAs in the probe is low. But these clones will strongly hybridize to the tester-specific subtracted probe, which is enriched in cDNAs that correspond to differentially expressed mRNAs, including low-abundance transcripts (e.g. *Figure 6*, clones A10 and A11). Usually, 20–50% more clones that correspond to differentially expressed mRNAs can be revealed using subtracted probes, in addition to those that have been detected using un-subtracted probes (*Table 1*). Thus, differential screening using only unsubtracted probes (33, 34) can lead to loss of the rarest differentially expressed cDNAs. In addition, using subtracted cDNA probes provides additional confirmation of the results obtained using unsubtracted probes.

However, using subtracted probes has one serious disadvantage. It reveals all cDNAs that were enriched during subtraction. In most cases, only differentially expressed cDNAs will be enriched during the SSH procedure, but sometimes non-differentially expressed cDNAs may also be enriched. The percentage of cDNAs that are enriched in a subtracted library, but are not truly differentially expressed, increases with each PCR cycle performed to obtain the subtracted cDNA pool. Therefore, it is important to use the minimum number of PCR cycles during primary and secondary PCR (see *Protocols 8* and *9*). Furthermore, the percentage of such clones increases in subtracted libraries as the efficiency of subtraction decreases. In this case, differential screening using subtracted probes is not a reliable method for identifying clones of differentially expressed cDNAs. Thus, it is important to verify clones selected from differential screening by other methods, such as Northern blots, Virtual Northern blots or RT–PCR. For example, 95% of the clones detected by subtracted probes in liver-, lung-, and breast-specific libraries were confirmed by Virtual Northern blot analysis. But in the prostate-specific libraries, none of the clones detected by subtracted probes were confirmed.

Furthermore, for high-throughput analysis, it is reasonable to limit differential screening by using only non-subtracted cDNA probes. In this case, confirming differential expression of the candidate clones is probably unnecessary (33, 34), so the researcher may proceed directly to sequence analysis of the subtracted libraries.

5.2 Sequence analysis of subtracted libraries

Large-scale sequencing of differentially expressed clones from SSH-subtracted libraries in the study described here (*Table 2*), as well as from previous studies (6, 30, 32, 33), has shown that they are very complex in terms of the number of independent sequences. SSH-subtracted libraries are a very rich source of new sequences (*Table 2*) that either share partial homology to known genes (17 of 83) or lack any homology to known genes (66 of 83). Thirty per cent of new sequences in the tumour- and normal tissue-specific subtracted libraries generated in the present study (*Table 2*) is consistent with the average percentage of new sequences in SSH-subtracted libraries reported by others (30, 32, 33). Moreover, a significant number of differentially expressed mRNAs have homology with ESTs or genomic sequences (52 of 174). Usually, there is no information about the function of these sequences.

More than half of the genes with known functions have been previously reported to be associated with cancer (70 of 132), providing good evidence that these tumour-specific libraries are significantly enriched for known and unknown genes involved in tumour development. However, it should be pointed out that in our experiments, SSH was performed with tumour and normal tissues from only one patient in each case. It would be premature to speculate that the same sets of genes are differentially expressed in other patients afflicted with similar tumours.

It is usually necessary to analyse 300–500 clones from the subtracted library to be sure that you do not lose genes representing low-abundance transcripts (present study, 32, 33). Sequence data from this and other studies indicate that the majority of the clones will be picked two to six times, suggesting that there is a small degree of redundancy (present study, 30, 33). This finding confirms the high level of normalization of SSH libraries, which suggests that our subtracted libraries contain high- as well as low-abundance differentially expressed cDNAs. The normalization of the subtracted libraries is also confirmed by Northern blot analysis of differentially expressed cDNAs isolated by SSH. The exposure time for these clones varies from 30 min to several days (present study, 30, 33).

Nevertheless, some of the sequences are much more prevalent in subtracted libraries than are others (*Table 1*) (32). This reflects the fact that different cDNA fragments have different abilities to be enriched during the SSH procedure.

5.3 Efficiency of enrichment by SSH

Theoretically, the SSH method can result in up to a 10 000-fold enrichment of differentially expressed cDNAs (7). In our model system, the SSH technique enriched for rare sequences over 1000-fold in one round of subtractive hybridization (6, 7). Many examples of efficient enrichment for known differentially expressed cDNAs have been shown previously in several experimental systems (7, 33) with greater than 200-fold enrichments. Nevertheless, in practice, not all differentially expressed genes are equally enriched by SSH. The level of enrichment of a particular cDNA depends greatly on its original abundance, the ratio of its concentration in tester and driver, and the number of other differentially expressed genes. cDNA fragments from a gene that is upregulated 100-fold in the tester will be more efficiently enriched than fragments of a gene upregulated fivefold. Furthermore, the ability of some cDNA fragments to be rehybridized and/or PCR-amplified more efficiently than others has an effect on enrichment in SSH. Thus, some fragments of a given differentially expressed gene may be eliminated during the SSH procedure, whereas other fragments of the same gene may be enriched and isolated. *Table 1* shows examples of cDNAs that are highly prevalent in different subtracted libraries. In the liver-metastasized colon carcinoma-specific library, non-specific cross-reacting antigen was highly prevalent compared with all other redundant genes (*Table 1*). Most of the cDNA fragments corresponding to this gene had identical sequences, suggesting that this fragment is more efficiently hybridized and/or enriched than other fragments of the same gene. Fortunately, once even a small cDNA fragment is cloned and sequenced, the corresponding full-length cDNA can be quickly obtained using numerous approaches, including several PCR-based methods (35).

5.4 Quality of the starting sample

Starting samples that are highly complex, in terms of both the different number of cell types and the percentage of target cells in the entire cell population, can cause problems in performing successful SSH. When a tissue consists of many different cell types, RNA purified from this tissue contains an increased number of independent mRNA species. Such a highly complex poly(A)$^+$ RNA sample decreases the efficiency of SSH. For instance, prostate tissue contains a significantly less homogeneous population of cell types in comparison with other tissues used in our subtractions. Moreover, the prostate tumour sample consisted of only 30% tumour cells, whereas lung tumour, liver-metastasized colon carcinoma, and breast tumour samples consisted of 90% tumour cells. We believe that these factors contributed to the poor efficiency of the prostate-specific subtractions. Differential screening analysis of prostate tumour-specific clones showed that only 13% of the cDNA clones were upregulated, compared with 50–70% upregulated clones in the lung tumour-, breast tumour-, and liver-metastasized colon carcinoma-specific libraries (*Table 1*). For optimal results using SSH, the starting tissue should have the most homogeneous cell population possible. Homogeneity can be achieved by dissection of fixed or frozen tissues (36) and cell sorting (37).

It should be noted here that the high complexity of genomic DNA prevents the use of SSH for genomic DNA subtraction. In representational difference analysis (RDA) (38), this problem is solved by repeated rounds of hybridization and PCR. Because of the normalization step in SSH, subtractive hybridization cannot be repeated; however, the normalization step is worthwhile because it results in very high enrichment of rare differentially expressed genes and high complexity of the subtracted cDNA pool.

5.5 Complexity of the subtracted library

In general, the complexity of SSH-subtracted libraries is very high in comparison with libraries generated using other subtraction techniques. For example, in cDNA RDA, repeated hybridization cycles efficiently enrich the most abundant differentially expressed genes between tester and driver in each round, so that the final subtracted cDNA population consists of only a few sequences (5, 39). In contrast, SSH generates a more complex population of subtracted cDNA fragments, and the final subtracted pool of cDNAs appears as a smear with or without a few bands on a 2% agarose/EtBr gel (*Figure 3*). It has been confirmed by our sequencing analysis in this study (*Table 2*) and in previous ones (6, 30, 33) that this smear reflects the high number of independent clones in the subtracted library. We advise against cutting bands directly from the gel for further analysis (31); rather, we recommend that you clone the whole subtracted cDNA population (see Section 3.2.8).

The exact number of independent clones in each SSH subtracted library depends on the nature of the tester and driver mRNA samples. When the differences between tester and driver poly(A)$^+$ RNA populations are expected to be very high, the complexity of the subtracted library is significantly higher (present study, 6, 30) than in the case when tester and driver RNA populations are very similar (24). So in the present study, the biggest difference between tester and driver RNA populations would be expected for the subtraction of liver-metastasized colon carcinoma against normal liver tissue, because in this case, tumour and normal tissues were derived from different organs. Indeed, the percentage of differentially expressed clones in the liver-metastasized colon carcinoma-specific library is the highest in comparison with other subtracted libraries (*Table 1*).

However, we should stress that the background of non-differentially expressed cDNAs in SSH subtracted libraries also has a high complexity. Therefore, subtracted cDNA may appear as a smear due to the high background.

5.6 Ratio of driver/tester and hybridization time

The stringency of subtraction can be altered by changing the ratio of the driver/tester. Increasing the driver/tester ratio allows preferential enrichment of those genes that are highly upregulated in tester compared with driver and also decreases the background of non-differentially expressed clones in the subtracted library. Nevertheless, cDNAs with smaller differences in expression will be lost during the subtraction as the driver/tester ratio is increased. Likewise, decreasing the ratio allows the identification of those genes that are only slightly upregulated in tester compared with driver; however, a low driver/tester ratio will also increase the occurrence of false positive clones.

Decreasing the length of both the first and second hybridization steps results in recovery of high-abundance differentially expressed genes with a low background of non-differentially expressed cDNAs. However, this also decreases the probability of obtaining low-abundance differentially expressed genes. Moreover, if the tester does not contain high-abundance differentially expressed genes, decreasing the hybridization time will lead to a higher background. Similarly, increasing the driver/tester ratio will lead to a higher background if tester and driver lack the mRNAs with significant difference in transcription level.

5.7 Summary

All of these factors explain why the optimum driver/tester ratio and hybridization time depend on each particular pair of driver and tester used. As we were interested in the recovery of low-abundance differentially expressed genes, we used the maximum hybridization time, which is 8–12 h for the first hybridization and 14–16 h for the second hybridization. Although the best driver/tester ratio was 50:1 for most of the samples used in this study, we recommend the use of a 30:1 ratio, which consistently gives good results. In general, performing SSH with several different ratios and hybridization times side by side may help to choose the best hybridization conditions, which will result in subtracted libraries with the highest percentages of differentially expressed clones. However, usually no more than two times improvement can be achieved.

6 Conclusions

SSH combines a high subtraction efficiency with an equalized representation of differentially expressed sequences. In combination with high-throughput differential screening, SSH allows the rapid identification of differentially expressed genes while reducing the number of false positive clones. Virtual Northern blot analysis can be used as a general tool for confirmation of differential gene expression. Our results demonstrate that SSH generates highly complex subtracted libraries that contain high-abundance as well as low-abundance differentially expressed cDNAs. The SSH technique should be applicable to many other studies requiring the enrichment of cDNAs derived from differentially expressed genes of a particular tissue or cell type.

Acknowledgements

We thank Dr A. Gudkov for critical reading of the manuscript and Jeff Baughn for preparing the figures for this chapter.

References

1. Hedrick, S. M., Cohen, D. I., Nielsen, E. A., and Davis, M. M. (1984). *Nature*, **308**, 149.
2. Sargent, T. D. and Dawid, I. B. (1983). *Science*, **222**, 135.
3. Hara, E., Kato T., Nakada, S., Sekiya, S., and Oda, K. (1991). *Nucleic Acids Res.*, **19**, 7097.
4. Wang, Z. and Brown, D. D. (1991). *Proc. Natl Acad. Sci. USA*, **88**, 11505.
5. Hubank, M. and Schatz, D. G. (1994). *Nucleic Acids Res.*, **22**, 5640.
6. Diatchenko, L., Lau, Y.-F. C., Campbell, A., Chenchik, A., Moqadam, F., Huang, B., Lukyanov, S., Lukyanov, K., *et al.* (1996). *Proc. Natl Acad. Sci. USA*, **93**, 6025.
7. Gurskaya, N. G., Diatchenko, L., Chenchik, A., Siebert, P. D., Khaspekov, G. L., Lukyanov, K. A., *et al.* (1996). *Anal. Biochem.*, **240**, 90.
8. Jin, H., Cheng, X., Diatchenko, L., Siebert, P. D. , and Huang, C. C. (1997). *BioTechniques*, **23**, 1084.
9. Gubler, U. and Hoffman B. J. (1983). *Gene*, **25**, 263.
10. Britten, R. J. and Davidson, E., H. (1985). In *Nucleic acid hybridization—a practical approach* (ed. Hames, B. D. and Higgins, S.), p. 3. IRL Press, Oxford.
11. Lukyanov, K. A., Launer, G. A., Tarabykin, V. S., Zaraisky, A. G., and Lukyanov, S. A. (1995). *Anal. Biochem.*, **229**, 198.
12. Siebert, P. D., Chenchik, A., Kellogg, D. E., Lukyanov, K. A., and Lukyanov, S. A. (1995). *Nucleic Acids Res.*, **23**, 1087.
13. Kellogg, D. E., Rybalkin, I., Chen, S., Mukhamedova, N., Vlasik, T., Siebert, P. D., and Chenchik, A. (1994). *BioTechniques*, **16**, 1134.
14. Chenchik, A., Zhu, Y. Y., Diatchenko, L., Li, R., Hill, J., and Siebert, P. D. (1998). In *Gene cloning and analysis by RT–PCR*, (ed. Siebert, P. D., and Larrick, J. W.), p. 305. Molecular Laboratory Methods Number 1. Biotechniques Books (Division of Eaton Publishing).
15. Hunt, T. and Scherr, C. J. (1994). *Curr. Opin. Cell Biol.*, **6**, 833.
16. Hirschi, K. K., Xu, C., Tsukamoto, T., and Sager, R. (1996). *Cell Growth Differ.*, **7**, 861.
17. Loewenstein, W. R. and Rose, B. (1992). *Semin. Cell Biol.*, 3, 59.
18. Giancotti, F. G. and Ruoslahti, E. (1990). *Cell*, **60**, 849.
19. Akiyama, S. K., Olden, K., and Yamada, K. M. (1996). *Cancer Metastasis Rev.*, **14**, 173.
20. Muller, B. M., Yu, Y. B., and Lang, W. E. (1995). *Proc. Natl Acad. Sci.*, USA, **92**, 205.
21. Stetler-Stevenson, W.G., Aznavoorian, S., and Liotta, L. A. (1993). *Annu. Rev. Cell Biol.*, **9**, 541.
22. Campisi, J. (1994). *Cold Spring Harbor Symp. Quant. Biol.*, **59**, 67.
23. Swisshelm, K., Ryan, K., and Sager, R. (1995). *Proc. Natl Acad. Sci. USA*, **92**, 4472.

24. Diatchenko, L., Chenchik, A., and Siebert, P. D. (1998). In *Gene cloning and analysis by RT–PCR* (ed. Siebert, P. D. and Larrick, J. W.), p. 213. Molecular Laboratory Methods Number 1. Biotechniques Books (Division of Eaton Publishing).
25. Mitchell, E. P. (1998). *Semin Oncol.*, Oct; 25 (**5** suppl. 11), 12.
26. Moscow, J. A., Schmidt, L., Ingram, D. T., Gnarra, J., Johnson, B., and Cowan, K. H. (1994). *Carcinogenesis*, **15(**12), 2769.
27. Daher, R. and Beaini, M. (1998). *Clin. Chem. Lab. Med.*, **36** (9), 671.
28. Diatchenko, L., Lukyanov, S., Lau, Y.-F. C., and Siebert P. D (1999). In *cDNA preparation and Characterization*. In *Methods in Enzymology*, eds Sherman M. Weissman. Academic Press, San Diego, CA, **303**, 349–380.
29. Bogdanova, E., Matz, M., Tarabykin, V., Usman, N., Shagin, D., Zaraisky, A., and Lukyanov, S. (1998). *Dev. Biol.*, **194**, 172.
30. Kuang, W. W., Thompson, D. A., Hoch, R. V., and Weigel, R. J. (1998). *Nucleic Acids Res.*, **26** (4), 1116.
31. Nemeth, E., Bole-Feysot, C., and Tashima, L. S. (1998). *J. Mol. Endrocrinol.*, **20**, 151.
32. Mueller, C. G. F., Rissoan, M.-C., Salinas, B., Ait-Yahia, S., Raval, O., Bridon, J.-M., Briere, F., Lebecque, S., and Liu, Y.-J. (1997). *J. Exp. Med.*, **186** (5), 655.
33. Von Stein, O. D., Thies, W. -G., and Hoffmann, M. (1997). Nucleic Acids Res., **25** (13), 2598.
34. Wong, B. R., Rho, J., Arron, J., Robinson, E., Orlinick, J., Chao, M., Kalachikov, S., Cayani, E., Bartlett III, F. S., Frankel, W. N., Lee, S. Y., and Choi, Y. (1997). *J. Biol. Chem.*, **272** (40), 25190.
35. Chenchik, A., Moqadam, L., and Siebert, P. D. (1996). In *A laboratory guide to RNA: isolation, analysis, and synthesis* (ed. Krie, P.), p. 273. Wiley, New York.
36. Jensen, R. A., Page, D. L., and Holt, J. T. (1994). *Proc. Natl Acad. Sci. USA*, **91**, 9257.
37. O'Hare, M. J. (1991). *Differentiation*, **43**, 209.
38. Lisitsyn, N., Lisitsyn, N., and Wigler, M. (1993). *Science*, **259**, 946.
39. Chang, D. and Denny, C. (1998). In *Gene cloning and analysis by RT–PCR* (ed. Siebert, P. D., and Larrick, J. W.), p. 193. Molecular Laboratory Methods Number 1. Biotechniques Books (Division of Eaton Publishing).

Chapter 6

Gene expression analysis by cDNA microarrays

Spyro Mousses, Michael L. Bittner, Yidong Chen,
Ed R. Dougherty, Andreas Baxevanis,
Paul S. Meltzer and Jeffrey M. Trent
Cancer Genetics Branch, National Human Genorne Research Institutes, National Institutes of Health, Bethesda, MD, USA

1 Introduction

1.1 cDNA microarray technology

The Human Genome Project (1) has spurred the emergence of 'genome-wide' approaches to study gene function, regulation and interaction termed *functional genomics*. Comprehensive analysis of differential gene expression is becoming a cornerstone of functional genomics. To this end, several new technologies have been developed for high-throughput differential gene expression analysis (2–8). The deposition of cDNAs in an array format to filters or glass is a hybridization-based analytical tool at the forefront of this field. The parallel gene expression of arrayed cDNAs on glass and subsequent analysis by means of two-colour fluorescence was pioneered in the laboratory of Dr P.O. Brown at Stanford. The first proof of principle paper showed gene expression analysis for an array with 45 Arabidopsis genes (2). This approach has since grown rapidly to high density microarrays with tens of thousands of genes. Today, many laboratories around the world have established cDNA microarrays for a range of applications and found it to be very useful for simultaneously profiling mRNA levels for thousands of genes.

The transition of cDNA microarrays from a working prototype to a practical and reliable technology remains the focus of enormous effort from industry and academia. This has required the development and standardization of various hardware, analytical software, statistical methodology, biological resources, and biochemical methodologies. Though a dream for many laboratories, all of the required functionality essential to reliably use the technology to find changes in gene expression levels between tissues, which provide abundant RNA, has been achieved. The system's performance is currently being evaluated in model systems where observation of known changes and verification of newly observed changes will allow stringent characterization of the reliability with which such profiling can be carried out. The ultimate goal of this effort is to develop the ability to scan the mRNA expression state of all 80 000 human genes simultaneously, and to be able to correlate this information with the underlying biochemistry and cellular biology producing the observed biological state.

In this chapter, we present a practical description of cDNA microarray technology as it has been refined and applied in the Microarray Project[a] at the National Institutes of Health (NIH). Widespread dissemination of knowledge about cDNA microarrays and implementation of many cDNA microarrays facilities, for a variety applications, will lead to a rapid evolution of this new technology.

1.2 cDNA microarray overview

A cDNA microarray experiment can be divided into several stages as outlined in Plate 2. First, is the construction of very high density microarrays of specific and distinct DNA hybridization 'targets', each one representing a single gene, that are spotted in an arrayed format on a glass support. Multiple glass slides, each containing thousands of spots of DNA, can be synthesized and used in subsequent experiments. Then for each experiment, complex 'probes' are made which consist of a pool of fluorescently labelled cDNAs. This step begins with the extraction and preparation of mRNAs from two populations of cells. Then each of the two mRNA is reversed transcribed separately with the incorporation of different fluorescently tagged nucleotides producing two populations of differentially labelled cDNA probes. The two complex labelled probes are combined and are then simultaneously hybridized to the cDNA targets on the microarray. A device called the 'reader' is used to detect the resulting fluorescence of the hybridized probes on the microarray after laser excitation. If a particular gene's mRNA predominates in one of the two samples, more of the corresponding cDNA will hybridize to the spot representing that gene and the colour of the fluor it is tagged with will predominate. The ratio of the intensity of fluorescence from the two labelled cDNAs on a particular spot is determined using the reader's confocal laser scanning microscope. The resulting data can then be put into a database and analysed. Various bioinformatics approaches have been developed to process and visualize the enormous quantity of data generated and further approaches are being developed to compare gene expression profiles from multiple experiments.

[a] The Microarray Project is a collaborative research effort between numerous intramural scientists in multiple institutes and divisions of the National Institutes of Health (NIH), including the National Human Genome Research Institute (NHGRI), National Center for Biotechnology-Information (NCBI), National Cancer Institute (NCI), National Institute of Neurological Disorders and Stroke (NINDS), Biomedical Engineering and Instrumentation Program (BEIP), Division of Computer Research and Technology (DCRT) and many others. (www.nhgri.nih.gov)

2 Preparation and printing of target DNA

2.1 Target DNA considerations: source, type, and large-scale production

The choice of which genes and how many will be printed on a glass microarray depends on the availability of clones and sequences, the capacity of the slide, but ultimately on the purpose of the experiment. For the Microarray Project, we have utilized a '22K' human gene set comprised of from 22 320 human UniGene clusters, each of which represents a unique human gene (9). At least one sequence in each cluster corresponds to a physical cDNA expressed sequence tag (EST) clone (10, 11) that is available from the IMAGE consortium. Within the set are 4001 clones corresponding to known genes. All of the other clones represent unknown genes. It is this non-redundant 'backbone' of ESTs that our laboratory has used in selecting clones for arraying. The 22K set of clones has been re-arrayed from the original libraries by Research Genetics Inc. The 22K sequences represent a collection of 'potentially interesting' genes with which to test large-scale, high-throughput gene expression technologies. As the capacity of the microarrays and time it takes to print them does not permit us to print the entire 22K set, we make subset microarrays. The choice of which of these clones to be put on the microarray depends largely on what it will be used for. In some cases, cDNA microarrays are used for the profiling of known genes, while in other cases unknown or novel genes are preferred on the microarrays to facilitate gene discovery. For the study of DNA copy number by microarrays (12), genomic mapping information for the arrayed clones is desirable.

Currently, in our laboratory, slides with 6000–8000 clones are routinely printed. Custom tissue-specific cDNA microarrays can also be constructed using mostly clones with ESTs expressed in libraries of a specific tissue. For most applications where the profiling of functionally relevant genes plus gene discovery are the goals, cDNA clones from a combination of named genes and ESTs are incorporated into the microarray.

2.2 Target DNA preparation

2.2.1 Miniprep protocol: isolation of clones from bacteria

The protocol below describes a high throughput method for obtaining plasmid DNA used for subsequent polymerase chain reaction (PCR) amplification of the clones. Typically, we use the 96-well plate format and process up to12 plates at the same time (1152 clones). The IMAGE consortium clones we use are mostly obtained from Research Genetics Inc. and are provided as *Escherichia coli* cultures in 96-well plates. To isolate the plasmid DNA we use the Edge BioSystems (Gaithersberg, MD) 96-well alkaline lysis Miniprep kit (cat. 91528). The Miniprep DNA is then used as template for PCR amplification of the cDNA inserts and followed by purification of the PCR product.

Protocol 1

Pre-growth of clones to ensure maximum plasmid production

Equipment and reagents

- LB broth containing ampicillin
- 200 proof ethyl alcohol
- 'Zip-lock' bag containing a moistened paper towel

Method

1. Add 100 μl LB broth containing 200 $\mu g\,ml^{-1}$ ampicillin to each well of a number of 96-well plates (round bottom). (We use the 850 μl 12-channel head on the Matrix Electrapette to draw 800 μl and deliver 100 μl eight times.)
2. Thaw frozen 96-well library plates containing bacterial cultures and spin briefly, 2 min at 1000 r.p.m., to remove condensation and droplets from the sealer.
3. Sterilize the inoculation pins between samples by partially filling a container with 200 proof ethyl alcohol and dip the 96-pin inoculating block in the alcohol. Then flame the pins, using a lit gas burner as an ignition source. The alcohol container should be flame proof and you have a tight fitting lid at hand to cap the container with in case of fire. *Caution: Avoid the danger of fire by making certain that the flames are out before returning the inoculation block to the alcohol bath.*
4. After briefly allowing the inoculation block to cool, dip the pins in the library plate, then inoculate the equivalent LB plate ensuring correct orientation. Re-flame the inoculation block (as in step 3) and return it to the ethyl alcohol bath, *after flame is extinguished.*
5. Repeat as necessary for each plate you need to inoculate.
6. Reseal the library plates. Store the library plates at −70 °C.
7. Place inoculated LB plates with lids into a 'zip-lock' bag containing a moistened paper towel. Place the bag containing plates in a 37 °C incubator overnight.

Protocol 2

Inoculating deep well blocks

Equipment and reagents

- 96 Well Culture Block (P/N 4050066, Edge BioSystems 96-well alkaline lysis Miniprep kit)
- Superbroth containing ampicillin

Method

1. Fill each well of a 96 Well Culture Block (96-well alkaline lysis Miniprep kit) with 1 ml of Superbroth containing 200 $\mu g\,ml^{-1}$ ampicillin. (We use the Matrix eight channel 1250 μl Impact2 pipette to add 1000 μl to each well.)

Protocol 2 continued

2 Using the 96-pin inoculating block, as above, inoculate the 96 Well Culture Blocks.
3 Place blocks with lids taped in place, in the 37 °C shaker incubator with shaking (200 r.p.m.), for 24 h.

Protocol 3

Isolation of plasmid DNA

1 Isolate plasmid DNA from the cultures using Edge BioSystems 96-well alkaline lysis Miniprep kit (cat. no. 91528) according to the manufacturer's protocol.
2 The DNA is then stored at −20 °C and used as template for PCR amplification (see *Protocol* 4).

2.2.2 Protocol for PCR amplification of clones

Image consortium clone DNA isolated as described in Section 2.2.1 is used as a template for PCR amplification with M13 primers. Again, we use the 96-well format and typically amplify templates 12 plates at a time.

Protocol 4

PCR reaction

Equipment and reagents

- 96-well PCR plates and sealers are from Robbins Scientific
- Cycleplate TM, Thin wall PCR plate, 10 plates/case, part no. 1038-00-0
- Cycleseal TM, PCR Plate Sealer, part no. 1038-00-0
- AEK M13 forward (F) and reverse (R) primers, a custom oligo synthesized by Midland Certified
- We use the MJ Research (DNA Engine Tetrad) PTC-225 Peltier Thermal Cycler
- Perkin Elmer 10 × PCR Buffer, part no. N808-0189
- Perkin Elmer AmpliTaq part no. N808-4015
- Each of four dNTPs (100 mM stocks) from Pharmacia (no. 27-2035-02)

Reagents

- AEK M13 F 5′-GTTGTAAAACGACGGCCA-GTG-3′ stock concentration 1 mM
- AEK M13R 5′-CACACAGGAAACAGCTATG-3′ stock concentration 1 mM

Method

1 A PCR reaction mix is made by combining the following reagent:[a]

Protocol 4 continued

Reagent	[Stock]	[Final]	Volume per reaction
PCR buffer	10×	1×	10 μl
dATP	100 mM	0.2 mM	0.2 μl
dTTP	100 mM	0.2 mM	0.2 μl
dGTP	100 mM	0.2 mM	0.2 μl
dCTP	100 mM	0.2 mM	0.2 μl
AEK M13 for.	1000 μM	0.5 μM	0.05 μl
AEK M13 rev.	1000 μM	0.5 μM	0.05 μl
Ampli Taq Pol.	5 U μl^{-1}	0.05 U μl^{-1}	1 μl
dH_2O			87.1 μl

2 Using a multichannel pipette, transfer 99 μl of the master mix to each well of disposable thin wall PCR plate (label plates to match the template plate labels).

3 Using a multichannel pipette, transfer 1 μl of appropriate template DNA in each well taking care to keep the plate orientation and order.

4 Cover the plates with Cycleseal PCR plate sealers and place in thermocycling device. We use the MJ Research DNA Engine Tetrads because they each have four 96-well blocks and we use three machines to amplify 12 plates simultaneously.

5 Amplify the templates using the following cycle conditions:

Step	Temperature	Time
1	96°C	30 s
2	94°C	30 s
3	55°C	30 s
4	72°C	150 s
5	Repeat steps 2–4→25 times/cycles	
6	72°C	5min→END

6 Each PCR product can be analysed by electrophoresis of 2 μl of product on a 2% agarose gel containing 0.5 $\mu g\,ml^{-1}$ ethidium bromide, using 1 × TAE as a running buffer. We obtain a digital image of the gel under ultraviolet (UV) illumination and analyse electrophoresis products to ensure that a single band of distinct size is produced for each sample. The intensity of the band gives an estimate of the relative amount of product.

7 Quantification of PCR products can also be accomplished by fluorimetry. We add 100 μl of dilute Pico Green solution to each well of a Microfluor W 'U' Bottom 96-well plate. Then, 5 μl of dilute (1:120) PCR product is added to each plate. The fluorescence is then measured in a Perkin Elmer Luminescence Spectrometer LS50B.

[a] We typically scale this reaction up by about 1250-fold and make a master about mix to accommodate 12 plates (1152 reactions) plus a small surplus to accommodate for transfer loss.

2.2.3 PCR product purification protocol

Protocol 5

PCR product purification protocol

Equipment and reagents

- Ethanol/acetate precipitation mix: 150 mM sodium acetate (pH 6) in ethanol
- SSC

Method

1. Prepare an ethanol/acetate precipitation mix (150 mM sodium acetate (pH 6) in ethanol). Add 200 μl of the ethanol/acetate mix to each well of a conical bottom 96-well plate.
2. Using multichannel pipettor, transfer the remaining (approximately 98 μl) PCR product to the equivalent well containing ethanol/acetate. We use a multichannel pipettor, which draws, delivers and then mixes the PCR product with the ethanol/acetate.
3. Place plates in the −80 °C for a period of 1 h or overnight at −20 °C to precipitate.
4. Allow plates to thaw (to reduce brittleness and melt any ice) and spin in a Sorvall Super 21 (Sorvall Inc.). We typically spin four stacks of three plates at 3200 r.p.m. for 1 h (rubber pads between the stacked plates helps prevent cracking).
5. After centrifugation, the ethanol/water supernatant is removed and a 70% ethanol wash is added. We use a BioRad 1575 ImmunoWash station to remove ethanol and add 70% ethanol more efficiently.
6. Centrifuge the plates as in step 5 at 3200 r.p.m. for 1 h. Allow plates to dry overnight, covered with a clean paper towel in a drawer.
7. Re-suspend the PCR product by adding 40 μl of 3 × SSC to each well. Seal plates with foil sealer, making sure that all wells are tightly sealed. Place the plates in a 'zip lock' bag with a moist paper towel and place in a 65 °C oven for 2 h. Allow plates to cool slowly to prevent condensation on the sealer and upper rim of the well. Remove the cooled plates and store at −20 °C.

2.3 Printing of DNA target microarrays on glass slides

2.3.1 The arrayer

The printing process refers to the sequential transfer of individual purified PCR amplified fragments from a 96-well microtitre tray to an exact, predefined location on glass slides. The 'arrayer' encompasses various instruments that facilitates the high throughput process of printing DNA targets on to multiple glass slides at very high density. Specifically, the arrayer is composed of a robotic arm which moves (*x*, *y*, *z* axis) the 'quill pen probes' into position to either pick up or to spot down the DNA (sodium dodecyl sulphate (SDS) solution), a manifold

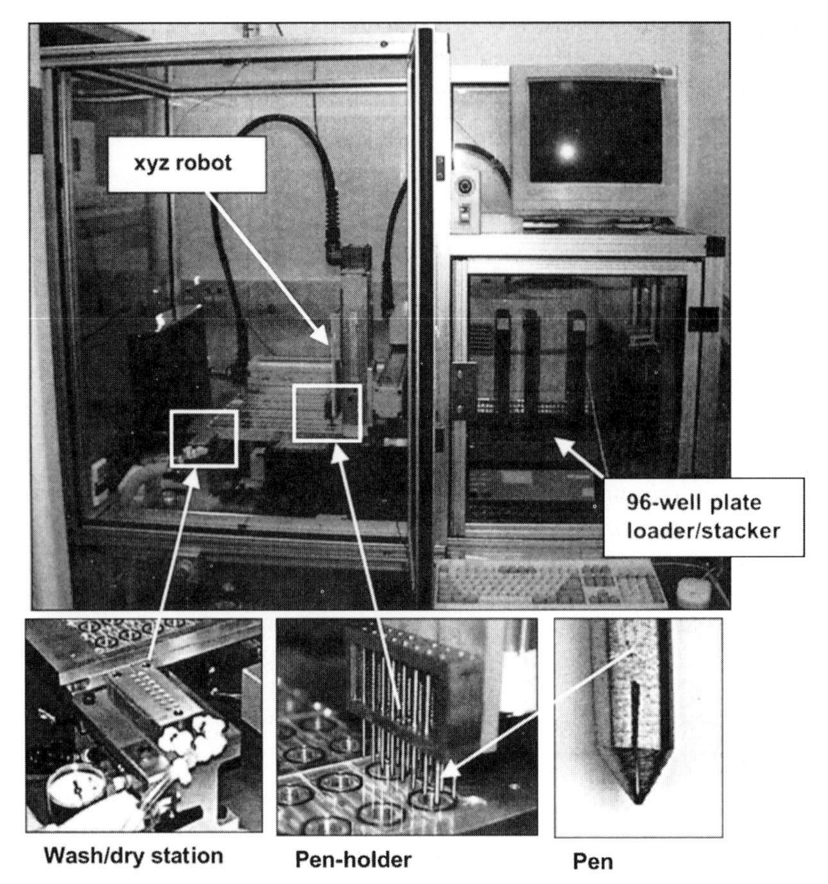

Figure 1 The arrayer printing layout. (A) The pens are attached to a bracket connected to a robotic arm, which manoeuvres them to various stations around the arrayer. (B) The 96-well tray station containing the DNA to be printed. (C) Glass slide station. After the pens dip into the DNA solution at the 96-well tray station they move to the glass slides, which are assembled on a platform and held in place by vacuum.

to hold the slides, a wash station to wash the 'quill pens' when a different DNA is to be picked up, a place to house the microtitre plates, an air flow cabinet to keep everything in, and a computer which orchestrates the operation of the components (*Figure 1*).

2.3.2 Essential components of the arrayer

DNA targets are picked up and spotted on the glass slides using a stainless steel probe which functions like a quill pen. It has a small slit in the middle and comes to a point allowing it to hold a very small volume of liquid. The robot moves the probe holder, with 16 stainless steel quill pens spaced on 9 mm centrelines to conform to the well spacing of standard 96-well microtitre trays. The probes are spring loaded to accommodate small differences in probe length and slide and well positions.

The vacuum chuck includes a removable platen so that the slides may be loaded and unloaded elsewhere, for example, in a clean hood. Each slide rests on two O-rings that seal it to the vacuum manifold. An external vacuum source can be connected to the manifold with a manual valve that permits releasing the vacuum for removal of the slides. Low ridges position the slides laterally.

The microtitre-tray stacker/loader holds twenty 96-well microtitre trays and presents them one at a time to the robot 'load' station when commanded by the computer. The wash/dry station flushes the probe tips with clean water and then dries them with a blast of clean air from the cabinet, pulled past the tips with a vacuum. The vacuum blast is of a greater flow than can be provided with a typical laboratory vacuum pump and is accomplished with a vacuum ballast tank, which is constantly pumped by a normal pump and then released only during the dry cycle.

A PC controls all of the other components and allows operator input of various parameters, such as number of probes, trays, and slides, spot spacing and pattern on the slides, duration of each cycle component and speeds and accelerations of the robot. A cabinet protects the operator and others in the lab from sudden movements of the robot and allows the maintenance of clean and humid (~65%RH) conditions around the trays and slides.

The air flow in the cabinet in the vicinity of the slides and trays is kept clean by a re-circulating blower, HEPA filter and ducting that keeps class 100 air circulating over the slides at about 100 cubic ft min^{-1}. Humidity is controlled with a room humidifier controlled by an industrial humidstat.

2.3.3 Arrayer performance

The arrayer is able to put down 16 spots on each of 48 slides, and to wash and dry the probes for the next set of cDNAs for the next tray, in about 70 s. Most of this time is taken up with the actual spotting, as the wash and dry cycles are about 2 s each and the loading is about 10 s. Thus, the contents of one 96-well tray can be spotted every 7 min. To produce 48 slides, each with 10 000 spots, takes about 12 h.

2.3.4 Treatment of glass slides

Treatment of slides with a coat of poly-L-lysine allows the target DNA to adhere to the surface and minimize loss during hybridization. The glass slides also need to be treated after the printing process. To reduce non-specific binding of strongly negatively charged probe on microarray slides, the positively charged amine groups on poly-L-lysine coated slides are passivated by reaction with succinic anhydride. The following two protocols have worked best in our hands for treating glass slides before and after printing.

Protocol 6

Poly-L-lysine slide coating protocol

Equipment and reagents

- Distilled, deionized water
- Ethanol (the source alcohol should be examined in a fluorometer to insure that it has very low levels of contaminating fluorescent organic compounds)
- Becton Dickinson, Gold Seal slides, no. 3011. (These slides have consistently low intrinsic fluorescence)
- Poly-L-lysine (pre-made 0.1% w/v solution from Sigma (P8920))

Important—Wear powder-free gloves at all times and avoid contact with detergents or other compounds that may cause background fluorescence.

Method

1. Place new Gold Seal microscope slides into a 50-slide rack (stainless steel/glass).
2. Prepare cleaning solution (enough for two 50 slide tanks): 400 ml H_2O; 100 g NaOH; 600 ml 95% ethanol. (Dissolve NaOH in water, then add ethanol. Stir until solution is clear. If the solution doesn't clear, add H_2O until it does.)
3. Submerge the rack in the cleaning solution (use glass tanks) and shake for 2 h.
4. Remove slides and rinse with water five times, 2–5 min each.
5. Move clean slides to a 25–30 slide rack (plastic or glass) so that they can be centrifuged after coating.
6. Prepare poly-L-lysine solution as follows: (enough for two boxes of 25 slides each): 35 ml poly-L-lysine (0.1% w/v) (Sigma P8920); 35 ml Tissue Culture phosphate-buffered saline (PBS); 280 ml H_2O.
7. Submerge rack in poly-L-lysine solution and shake for 1 h.
8. Rinse once in H_2O for 1 min.
9. Centrifuge rack in a low speed swinging holder centrifuge to remove free liquid.
10. Immediately transfer to a clean slide box.
11. Allow slides to age for 2 weeks before printing. Aged slides will be very hydrophobic (water drops leave no trail when they move across the surface).
12. Autoclave 20 min (for 1 litre).

Protocol 7

Blocking slides after printing with succinic anhydride

Reagents

- 1-Methyl-2-pyrrolidinone Aldrich no. 32 863–4
- Succinic anhydride Aldrich no. 23 969–0
- Borate buffer: made by adjusting the pH of boric acid (ACS grade) with sodium hydroxide (ACS grade)

Protocol 7 continued

Method

1 After printing DNA to the slides, the slides are allowed to stand for 1 week at ambient temperature in a closed slide box before UV cross-linking. A number of groups have found that rapid or slow hydration of the DNA on the slide after printing, followed by a quick drying step improves DNA distribution or signal strength. This has not been observed for materials prepared by our procedure, so it is routinely omitted. After standing for 1 week, slides are UV cross-linked. A dose of 450 mJ is applied with a Stratagene Stratalinker.

2 Cross-linked slides are put in an upright, stainless steel slide carrier, and the slide rack is placed in a clean glass tank. Dissolve 6 g of succinic anhydride in 325 ml of 1-methyl-2-pyrrolidinone,[a] then add 25 ml of 1 M sodium-borate buffer immediately before adding slides. (When water is added, the anhydride will begin to rapidly decompose, so add the mix to the slides very quickly.) Shake the slides for 20–30 min—some precipitation will occur. Prepare boiling water in a separate clean glass tank, so that it will be ready for after the reaction (in about 20 min).

3 Remove slide holder from the passivation reaction and dunk immediately in boiling water. Turn off heat source and let stand for 2 min in the nearly boiling water bath. After 2 min remove slide holder from boiling water and dunk in 95% or 100% ethanol.[b]

4 After 1 min in ethanol, remove the slides and centrifuge dry in a low speed swinging holder centrifuge. A centrifuge fitted for centrifuging 96-well plates usually will hold a 25–30-slide holder. Allow the slides to dry overnight in a dust-free cabinet and then rack in a clean slide box. (Simple plastic slide boxes, with no paper or cork to shed particles are preferred.)

[a] Caution: 1-Methyl-2-pyrrolidinone is a teratogen. Carry out all steps involving open containers of 1-Methyl-2-pyrrolidinone, in a fume hood.

[b] Ethanol is from a source tested to see that it has no inherent fluorescence.

3 Labelling and hybridization of complex cDNA probes to arrayed targets

3.1 RNA preparation

The quality of the RNA coming into the labelling will have a marked effect on the quality of the hybridization. RNA preparations, which look good by the standard molecular biology criteria, can give poor results. The amount of RNA required for each hybridization varies from about 40 to 200 μg for each of the two samples being compared. The exact amount should be determined empirically and depends on the size of the array, and the efficiency of labelling. In our hands, Cy5 labelling is less efficient than Cy3 labelling and requires more input RNA. We have tested extraction methods and found satisfactory results with a sonication in chaotrope/RNeasy (Qiagen)/Triazol extraction protocol as described below.

Protocol 8

Preferred RNA extraction protocol

Kits/reagents

- Virtis VirSonic 60 (or 100) with micro probe (conical titanium probe 1.8 mm diameter tip)
- Qiagen RNeasy Midi Kit no. 75142
- BRL Trizol Reagent no. 15596–018

Method

1. Cultured mammalian cells are grown in large plates (400 cm^2 plates) if adherent or roller bottles (850 cm^2 bottles) if non-adherent cell lines. (The Qiagen RNeasy Midi Kit protocol should be consulted to determine the amount of tissue or cultured cells of various types is required for specific applications.)
2. Harvest cells by scraping and centrifugation into 50 ml conical polypropyl tubes.
3. Centrifuge and remove growth media.
4. Wash once with 15 ml of PBS, centrifuge and remove wash.
5. Add 2 ml of Qiagen buffer RLT (with β-mercaptoethanol) per 3×10^7 cells and vortex vigorously.
6. Sonicate the very viscous lysate with short bursts to reduce viscosity. It should be possible to bring the lysate to low viscosity with two to three 5-s bursts.
7. (Virtis VirSonic 60 (or 100) with micro probe (conical titanium probe 1.8 mm diameter tip)) (With VirSonic 60, setting 2, dissipated power approximately 5–10 W.)
8. Chromatograph according to Qiagen kit instructions. Spin only in a room temperature centrifuge. Note: it is important *not* to spin in a refrigerated centrifuge.
9. At the final elution stage, elute with two successive water washes of 150 μl.
10. Extract the eluted RNA a second time by adding 1 ml of Trizol per 300 μl of eluant and vortexing.
11. Add 200 μl of chloroform per millilitre of Trizol to partition the phases and centrifuge to separate.
12. Recover the top aqueous layer and precipitate with 0.5 ml of isopropanol per ml of RNA-containing buffer.
13. Wash the pellet twice with 70% ethanol.
14. Resuspend (and pool if desired) the pellets in 200 μl of water.
15. Precipitate with 600 μl of ethanol and 20 μl of 3 M sodium acetate (pH 5.2).

3.2 Direct labelling of cDNA using fluorescent dyes

The labelling of complex probes is accomplished by direct incorporation of fluorescent nucleotides during a reverse transcription (RT) reaction. The quality of the RNA going into the RT reaction is critical. Currently the factors of labelling efficiency, fluorescent yield, spectral separation, and tendency toward non-specific binding make the Cy3/Cy5 pair the most useful for our detection system.

Protocol 9

Fluorescent labelling protocol: single round RT labelling from total RNA

Reagents

- Nucleotide Mix
- 10 × low T dNTPs (use 100 mM dNTPs from Pharmacia (27-2035-02))

dNTP	μl	mM final (1/10) concentration
dGTP	25	0.5
dATP	25	0.5
dCTP	25	0.5
dTTP	10	0.2
Water	415	-
Total volume	500	

- Fluorescent nucleotides from Perkin Elmer Applied Biosystems Division: PA 53022 FluoroLink Cy3-dUTP 1 mM; PA 55022 FluoroLink Cy5-dUTP 1 mM
- Oligo(dT)$_{12-18}$ primer from Pharmacia (27-7858-01)
- RT (SSII RT enzyme) is BRL SuperScript II (18064-014)
- 5 × buffer (supplied with enzyme)

Method

1. Total RNA is prepared from tissue or tissue culture resuspend the prepared RNA in a volume which will produce an RNA concentration of >6 mg ml^{-1} in DEPC water. If the RNA is recovered from a matrix as the final preparative step, and is still too dilute, concentrate as needed for labelling using a MicroCon30 (Amicon).
2. Determine the concentration of your RNA—read a small sample in 50 mM NaOH—assume 35 μg ml^{-1} for 1 A260 (10 mm path).
3. Normalization of the many factors that must accounted for in a two fluor experiment is now primarily achieved by reference to housekeeping genes distributed through the array. We still add a cocktail of synthetic cDNAs produced using phage RNA polymerases on cloned *E. coli* genes in the pSP64 poly(A) vector as a mass standard and RNA quality standard.
4. If possible, prepare a total RNA solution containing 100 μg of RNA in 17 μl of DEPC water if unavoidable, EtOH precipitate 100 mg of total RNA to concentrate sample for labelling reaction. *IMPORTANT: take care to remove all residual 70% ethanol wash either by air drying or vacuum—residual ethanol may seriously impede the efficiency of labelling.*
5. Resuspend the pellet in DEPC water to give a final volume of 17 μl.

Protocol 9 continued

6 Any residual precipitate in your reaction not wanted, so remove by centrifugation before proceeding.

7 Proceed with the labelling reaction as described below. Ensure that the RNA is added last. Fluor NTP RT labelling:

Component	**μl**
5 × first strand buffer	8
Oligo(dT)$_{12-18}$ (500 μg ml^{-1})	2
10 × low dT NTP mix	4
Fluor dUTP (1 mM)	4
0.1 M DTT	4
RNasin	1
Vortex and centrifuge briefly	
syn mRNA std (0.06 μg)	0.5
100 μg total RNA	17.0
Total	40

8 Vortex and centrifuge briefly Note: At all times minimize bubbles and foaming.

9 Hold reaction at 65 °C for 5 min, bring to 42 °C and add 2 μl of SSII RT enzyme and mix well into the reaction.

10 Incubate 42 °C for 25 min and then add 2 μl more the of SSII RT enzyme, again making sure that the enzyme is well mixed into the reaction solution.

11 Incubate 42 °C for 35 min and then add 5 μl of 500 mM ethylenediamine tetraacetic acid (EDTA), mixing well to ensure that the reaction has stopped.

12 Add 10 ml of 1 M NaOH and incubate at 65 °C for 60 min to hydrolyse residual RNA. Cool to room temperature and add 25 μl of 1 M Tris-HCl (pH 7.5).

Protocol 10

Probe cleanup and analysis

Equipment and reagents

- Cy5 probe
- 2% agarose gel in TAE
- TE: 10 mM Tris-HCl, 1 mM EDTA (pH 7.5)

Method

1 The labelled cDNA is purified by chromatograph on a microbiospin-6 Column (Tris) (Bio-Rad Laboratories) according to the manufacturer's instructions.

2 Take a 2 μl aliquot of Cy5 probe for analysis, leaving 17–18 μl for hybridization and run this probe on a 2% agarose gel in TAE (gel size is 6 cm wide × 8.5 cm long, 2 mm wide teeth). For maximal sensitivity when running samples on a gel for fluor analysis, use loading buffer with minimal dye and do add EtBr to the gel or running buffer.

Differential screening

DRG cDNA **Brain cDNA**

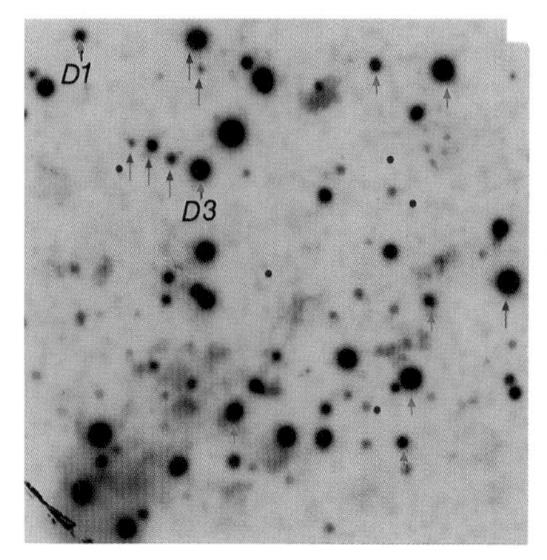

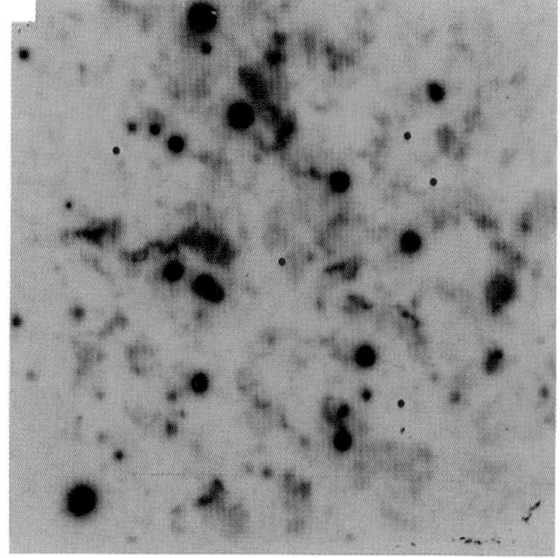

Confirmatory screening

Subtracted DRG cDNA **Brain cDNA**

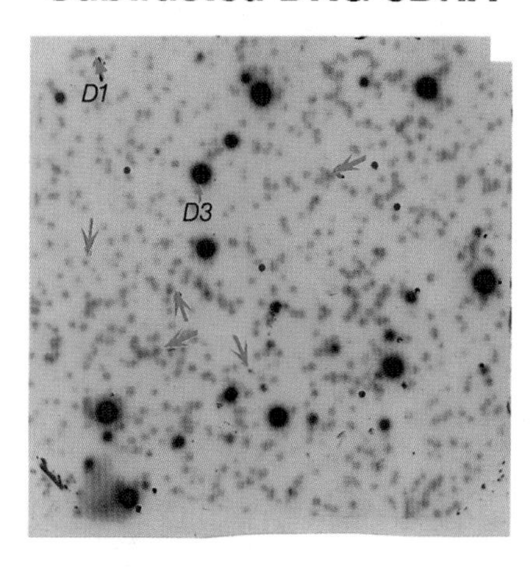

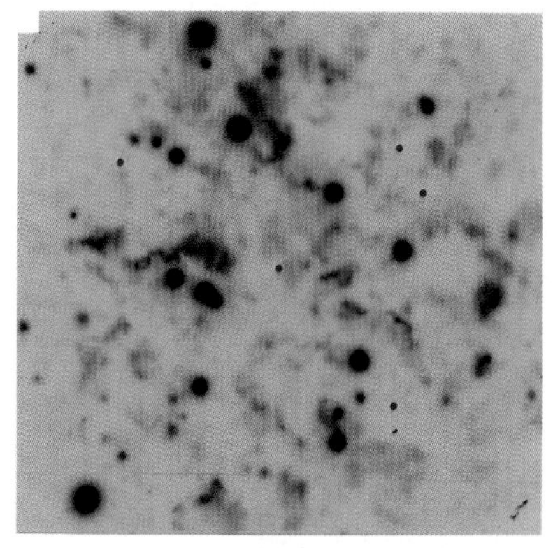

Plate 1. Representative autoradiographs (3-day exposure; final washing of the filters with 0.1xSSC, 0.5% SDS at 68°C) of the differential and confirmatory screening of subtractive cDNA library. Filters were probed with DRG [^{32}P] ss-cDNA (DRG cDNA), cortex plus cerebellum [^{32}P] ss-cDNA (brain cDNA), and also amplified subtracted random-primed [^{32}P] ds-cDNA (Subtracted DRG cDNA). Some DRG-specific transcripts are marked by green arrows and numbered on 'DRG cDNA' and 'Subtracted DRG cDNA' autoradiographs. Several house-keeping transcripts are indicated by red arrows.

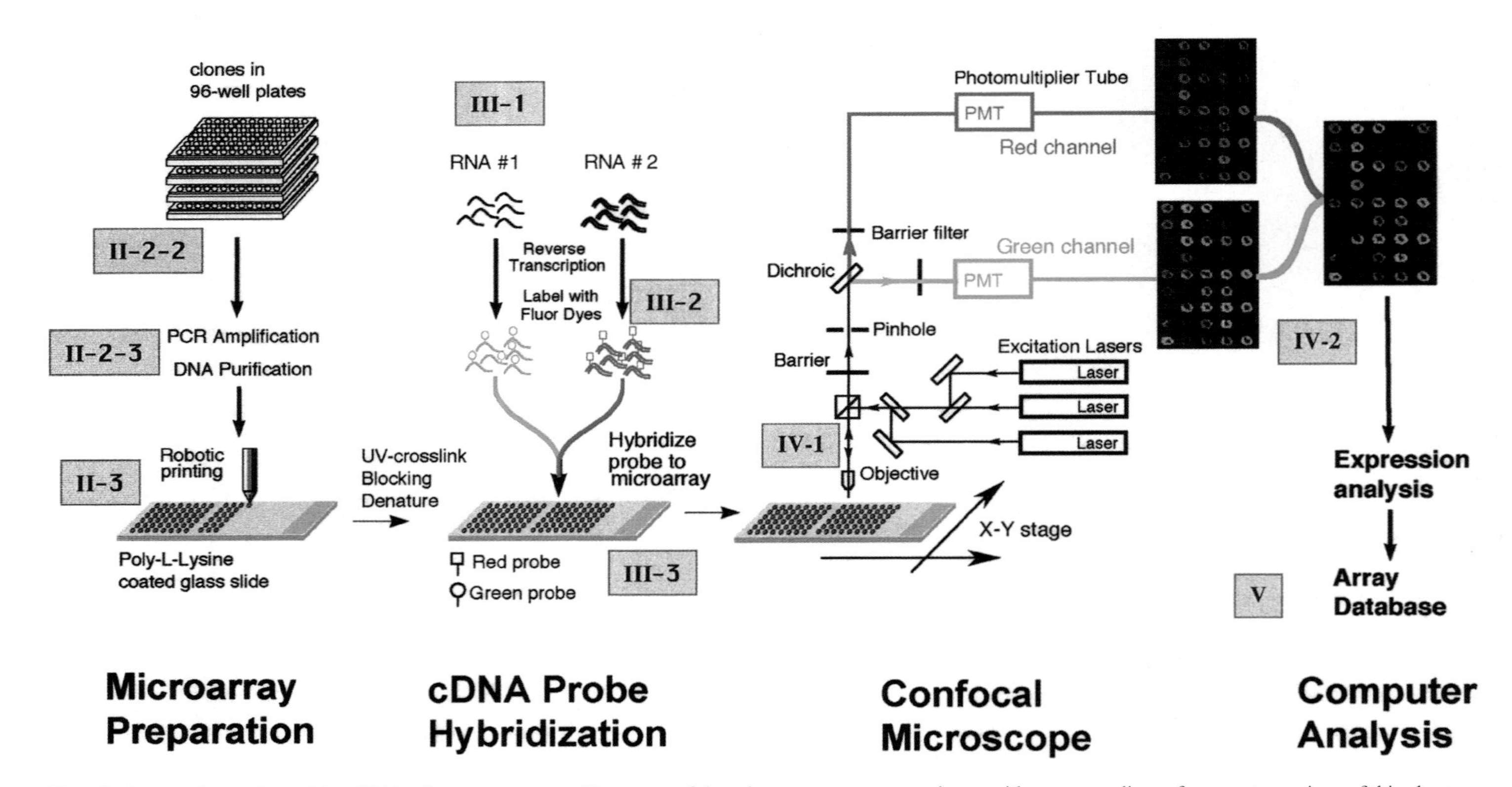

Plate 2: A general overview of the cDNA microarray process. The stages of the microarray process are shown with corresponding references to sections of this chapter. Adapted from Duggan *et al*. Nature genetics (19999) (Suppl. 1), pp 10–14.

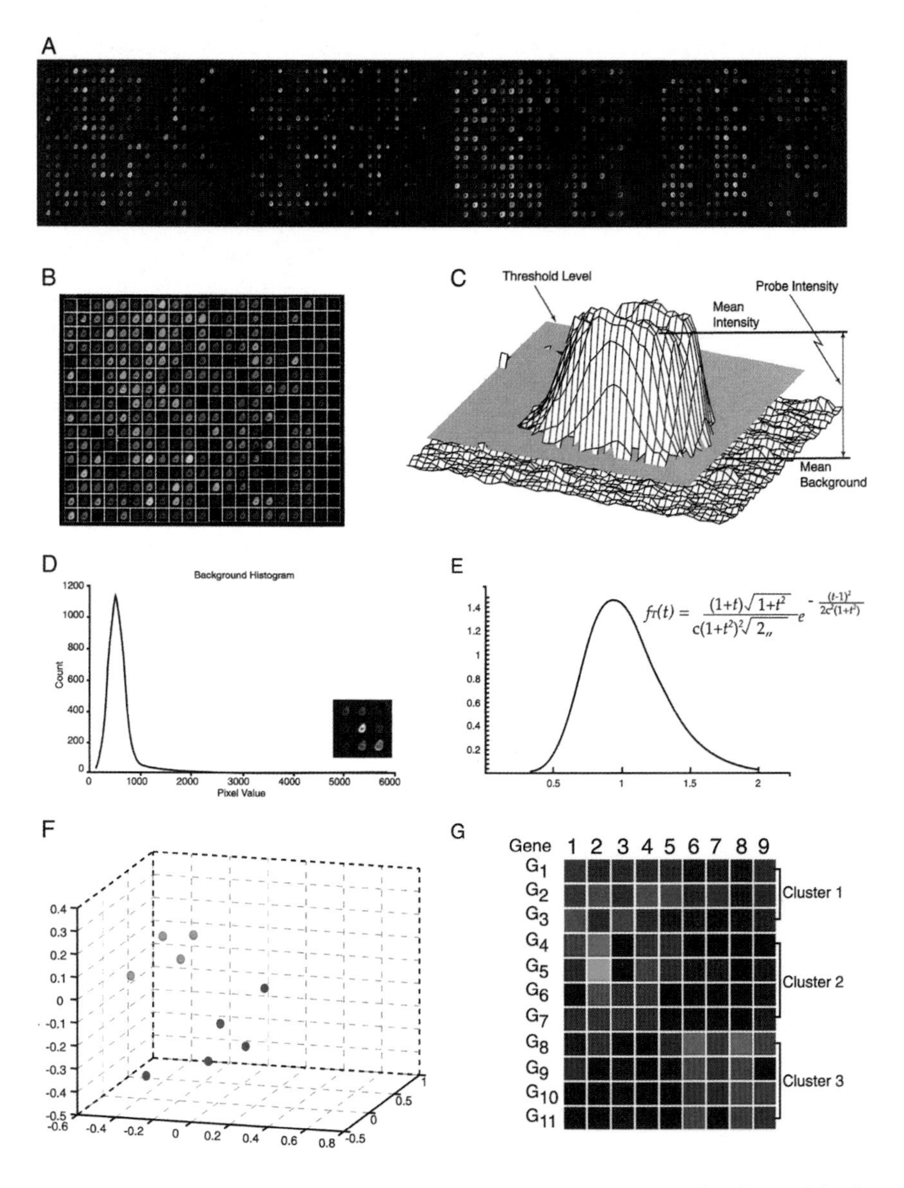

Plate 3: Image analysis. (A) A typical microarray image. (B) Microarray image aligned to predefined grid-overlay. (C) Target detection is showing sub-regions of intensity representing probe-hybridization-to-target area. The threshold surface is used to separate the target from background. The local background value is then subtracted from the reported probe intensities. (D) Location of mode of the background histogram provides the mean local background intensity. (E) The probability distribution of ratios obtained from expression arrays. (F) Multidimensional scaling allowing two clusters to be observed visually. (G) In this diagram the coloured square represents the expression level of a particular gene for each experiment. Cluster analysis allows related gene expression profiles to be placed together

Protocol 10 continued

3 Scan the gel on a Molecular Dynamics Storm fluorescence scanner (settings—red fluorescence, 200 μm resolution, 1000 V on PMT) successful labelling produces a dense smear of probe from 400 bp to >1000 bp, with little pile-up of low molecular weight transcripts (as in A) weak labelling and significant levels of low molecular weight material indicate a poor labelling (as in B). A fraction of the observed low molecular weight material is unincorporated fluor nucleotide. (Cy3 probe may be imaged on the Storm using the 488 laser; however, this provides inefficient excitation, and therefore dimmer images.)

4 To concentrate probes for hybridization, add 200 μl of TE (10 mM Tris-HCl, 1 mM EDTA, pH 7.5) to a microcon 30 membrane cartridge, then add the Cy5 and the Cy3 labelled probes together, pipette to mix, and concentrate to about 50 μl (approximately 3.5 min at 14 000 r.p.m. in Eppendorf 5415C).

5 Wash by adding 400 μl of TE (pH 7.5) and concentrating to about 20 μl (4–4.5 min at 14 000 r.p.m.).

6 Recover by inverting the concentrator over a clean collection tube and spinning for 3 min at 3000 r.p.m.

7 In some cases, the Cy5 probe will produce a gelatinous blue precipitate that is recovered in the concentrated volume. The presence of this material signals the presence of contaminants. The more extreme the contamination, the greater the fraction of the probe which will be captured in this gel. Even if heat solubilized, this material tends to produce uniform, non-specific binding to the DNA targets.

3.3 Hybridization to glass DNA microarrays

The volume required for the hybridization is dependent on the size of array used. For a typical microarray containing 5000–10 000 targets and using a standard 24 × 50 mm coverslip, a 30 μl hybridization volume is usually sufficient.

Protocol 11

Hybridization

Reagents

- poly(dA) Pharmacia (27-7988-01)—resuspend at 8 mg ml^{-1}
- yeast tRNA Sigma (R8759)—resuspend at 4 mg ml^{-1}
- Blocking species
- CoT1 DNA Life Technologies Inc.—concentrate 10-fold to 10 mg ml^{-1}

Method

1 Mix the following together:

Cy3 and Cy3 labelled and purified probe	~20 μl
Poly(dA) (40–60) (8 mg ml^{-1})	1 μl

Protocol 11 continued

Yeast tRNA (4 mg ml^{-1})	1 μl
CoT1 DNA (10 mg ml^{-1})	1 μl

2. Add 3 μl of 20 × SSC per 20 μl of hybridization mix volume. At this point you can optionally add 1 μl of 50 × Denhardt's blocking solution per 20 μl of hybridization mix (with a very clean probe the Denhardt's does not make any visible difference; however, it can reduce noise for less well behaved probes). Heat at 98 °C for 2 min, cool to 45 °C, and add 0.2 μl of 10% SDS per 20 μl of hybridization mix volume. Apply to microarray covered with a coverslip to hybridize (16–24) h at 65 °C in a sealed, humidified chamber.
3. Washing: Residual unbound probe is removed from the slides by washing 2–5 min each at room temperature in: first wash: 0.5 × SSC, 0.01% SDS; second wash: 0.06 × SSC. Note: Air drying of the slides after this step frequently leaves a fluorescent haze on the slide surface, so buffer is removed from the slides by a brief slow spin.
4. Place the slides in a slide holder and spin in a centrifuge equipped with a swinging carrier (horizontal) which can hold the slide holder (most centrifuges adaptable to centrifuging microtitre plates can be used for this purpose).
5. Scan slide in 'Reader' as soon as possible.

3.4 Critical parameters and troubleshooting

Typical problems include dispersed, fine-grain noise over the entire hybridized surface and non-specific binding of fluor to the zones of DNA immobilization on the slide. These problems seem likely to have some roots in contaminating carbohydrate, and as would be expected with carbohydrate, the problems are exacerbated by ethanol precipitation before and after labelling. Very impure preparations will frequently produce visible aggregates if precipitated after labelling and ethanol precipitation, which are essentially resistant to solubilization. It is well known that nucleic acids form strong aggregates with carbohydrate when either dried together or when co-precipitated. This interaction is the basis for nucleic acid immobilization on to chromatography supports such as cellulose. To minimize this sort of problem, we recommend preparative procedures that use few or preferably no ethanol precipitations during RNA preparation and labelling. In any method you choose, use at least the volumes of extractant and washing solutions suggested for the number of cells being processed. Appropriate or slightly excessive extraction/wash volumes tend to minimize noise.

4 Generation and analysis of image data

4.1 The reader

The reader designed at the NIH is based on the original design by Brown and colleagues, although a number of additional features have been added in-house.

Figure 2 The 'reader' built and used at the NHGRI is a computer-controlled inverted scanning fluorescent confocal microscope with a triple laser illumination system. For a more detailed description and specifications please refer to the microarray project web site at www.nhgri.nih.gov/DIR/LCG/15K/HTML/.

The reader is basically a computer-controlled inverted scanning fluorescent confocal microscope with a triple laser illumination system.

4.1.1 Essential components

The optical system is folded and arranged on an optical breadboard. The breadboard is hung with shock mounts in a vertical plane (to save space), inside a lightweight enclosure which also protects the optics from laboratory dust and personnel from laser light.

Illumination is from three air-cooled lasers: a 488 nm, 100 mW Argon ion laser for exciting fluorescein isothiocyanate; a 532 nm, 100 mW Nd:Yag for Cy3 and a 633 nm, 35 mW HeNe for Cy5. Any two lasers may be turned on simultaneously and their beams combined with dichroic mirrors and delivered to the specimen via a single dichroic and an objective lens (0.75 NA, 0.66 mm wd). The objective lens can be remotely and reproducibly focused with a digital controller.

The emitted light, after passing back through the objective and primary dichroic, is focused through a confocal pinhole and through a secondary dichroic on to two cooled PMTs which operate in parallel for the two different wavelengths.

The stage is a standard computer controlled microscope stage capable of 100 mm s^{-1} scans and 5 μm resolution. One or two standard 25 × 75 mm slides can be scanned at a time. Scanning is done in a comb pattern with data collected in both directions.

Data are acquired with a custom integrator and standard 16-bit A/D card in a 133 MHz Pentium. The operator can set the gain, speed, pixel size, pattern position, and size. All of the electronics and power supplies are mounted in the cabinet bay next to the optics.

4.1.2 Reader performance

At 100 mm s^{-1}, with 20-μm pixels, a 50 × 20 spot array with spots on 400-μm centres involves 400 traverses each about 20 mm long and can be scanned in about 4 min. We can reliably detect about 10 pg μl^{-1} of each species of cDNA. A 5000 spot microarray slide can usually be 'read' (scanned) in an hour or two.

4.2 Image data generation and analysis

The objective of the microarray image analysis is to extract probe intensities or ratios at each cDNA target location, and then cross-link printed clone information so that biologists can easily interpret the outcomes and perform further high-level analysis. However, the microarray image sources are not only from one print-mode (i.e. different printing tip arrangement or different arrayers (13) or one hybridization method (i.e. fluorescent (Stanford, NIH, etc.), radioactive probe, and others (14). Nevertheless, in order to simplify the presentation of information for this section, we have chosen to model the microarray images below from two fluorescent probes as our main processing input images. More information is available in (`http://www.nhgri.nih.gov/DIR/LCG/15K/HTML/`). A typical microarray image is shown in Plate 3(A).

For this particular image example, we hybridized an array of cDNAs with a Cy3 fluor-tagged sample and Cy5 fluor-tagged sample, and then individual fluorescent intensity images was combined by placing each image into the appropriate colour channel of the RGB colour image. In this experiment, there were a total of 1344 cDNA targets, printed from fourteen 96-well plates, using four print-tips. Given the print-mode of the arrayer, the software can easily track the cDNA targets on this and the other array, and map them back to their original 96-well plate position. In following sections, microarray image analysis is further divided into target segmentation and detection, background signal and fluorescent signal measurement, ratio analysis and last but not least the statistical gene expression clustering method.

4.2.1 Array target segmentation

As each element of an array is printed automatically to a pre-defined pattern and position, we can safely assume that the final probe signals from a regular array can be automatically aligned to a predefined grid-overlay. The initial position of the grid, as shown in the Plate 3(B), can be either (i) manually determined if no particular orientation markers are printed or no visible signals that can be used as orientation markers, or (ii) automatically determined if the orientation markers are presented in the final image and the entire array has no obvious missing row or column signals. Owing to the complication of the customized print procedures and various hybridization protocols, we assume the initial grid-overlay is manually determined. Usually, the initial target segmentation achieved by grid overlaying does not need to be precise. An automatic refinement of the grid position is preferable following the manual grid overlaying. The procedure we utilize is as follows: (i) detect strong targets; (ii) find their centres (e.g. centre of mass); (iii) regress four-corner coordinates of each sub-array from these centres of strong signals. This last regression step is important as it avoids the problem of false centres due to noise blobs or miscalculated centres. The final grid-overlay precisely segments each target, which enables future processing tasks to concentrate only one target. After target segmentation, the clone information can be attached to each target at this stage by deconvoluting the target position on the microarray slide back to its original 96-well plate position when the printing information is available on-line.

4.2.2 Target detection

One of the difficult image-processing tasks is to identify the target region within the bounding box. Each target is somewhat annular, shown in Plate 3(C), resulting from both how the robot print-tip places the cDNA on the slide and how the slide is treated. However, the final image of the target may simply be a collection of subregions within the nominal circular target region due to the variability introduced by cDNA deposition or the hybridization process. It is important that the final signal intensity be measured over regions corresponding to probe-hybridized-to-target area.

Conventionally a fixed thresholding method is used in target detection. The threshold value T is determined from the local background mean intensity μ_b and its standard deviation σ_b by the relationship $T = \mu_b + 3\sigma_b$. This concept is illustrated in Plate 3(C) where a flat threshold surface defined by T is used to separate target from background. However, the simple fixed thresholding method fails quite often due to variability of the background and the signal, particularly when the signal is weak (a frequent finding in cDNA array experiments). To avoid these problems, some sophisticated thresholding methods may be implemented. One of methods that we utilize is the Mann–Whitney method (15), which takes sample pixels from the background and then performs a rank-sum hypothesis test on the target pixels. The advantage of this method is to allow users specifying the confidence level of target detection in order to detect weak target.

4.2.3 Background intensity extraction

Typically, the background of the microarray image is not uniform over the entire array and, therefore, it is necessary to extract local background intensity.

The changes of fluorescent background across an array are usually gradual and smooth, and may be due to many technical reasons. Abrupt changes are rare; when they happen, the actual signal intensities of array elements near these changes are not reliable. Conventionally, pixels near the bounding box edge are taken to be background pixels, and thus the mean grey-level of these pixels provides an estimation of the local background intensity. This method become inaccurate when bonding box size is close to 10×10 pixels, or the target fills entire bounding box.

Fluorescent background is typically modelled by a Gaussian process (16). For example, if a larger area is chosen (e.g. a 40×40 box centred at a particular target) the grey-level histogram within the box is usually unimodal, as the majority of the background pixel values are similar, while the target pixel values spread up to very high grey levels. The location of the mode of the histogram, therefore, provides the mean local background intensity μ_b and the left tail of the histogram will provide the spread (standard deviation σ_b) of the background intensity. An example histogram is shown in *Plate 3(D)* where a typical unimodal histogram is derived from a region (the small insert at the right-hand side) with no specific efforts made to eliminate the target region, even though it contains a bright target.

4.2.4 Target intensity extraction

Intensity measurements are carried out after target regions are determined. For a two-colour system, we first unite the target region detected from the red channel and the target region detected from the green channel. The reason is simple: both probes were hybridized to the same target, so if we observe either one of them, the underlying region must belong to original target.

The probe intensity measurement is chosen to be the average grey level within the target region. Keeping in mind that the final measurement is the ratio of two intensities (R/G), the average measurement will provide, to some degree, a data smoothing effect. The local background value is then subtracted from the reported probe intensities from the red channel (R) and the green channel (G), as illustrated in *Plate 3(C)*; the ratio (R/G) is then calculated. Clearly, the ratio measurement is the ratio of two average intensity measurements. There are, of course, other choices for ratio measurement, including (i) the geometric average of ratios from every pixel location, and (ii) the linear regression slope of R/G grey values from every pixel location.

4.2.5 Ratio analysis

We have used expression ratio calculations to determine whether gene expression differs significantly between the red and green channels. Such an approach is intuitive because two similar samples lead to a R/G ratio close to 1. Assume the ratio (denoted as t) extracted from a microarray image satisfied the following conditions: (i) normality; (ii) independence; (iii) being sufficiently positive; and

(iv) having a constant coefficient of variation, c. We can approximate the ratio distribution as follows (15):

$$f_r(t) = \frac{(1 + t)\sqrt{1 + t^2}}{c(1 + t^2)^2\sqrt{2\pi}}\, e^{\frac{(1 - t^2)}{2c^2(1 + t^2)}}$$

Normally, ratios obtained from expression arrays possess the probability distribution depicted in Plate 3(E).

The distribution parameter can be estimated using a maximum likelihood method, and ratio calibration can be carried out by an iterative method (assuming that the two channels are not normalized). In order to satisfy the null hypothesis, which requires no intensity change from red to green, we have chosen a set of 'housekeeping' genes. These genes have been selected and experimentally verified as being stable in most experiments (R/G = 1.0). While being referred to as a 'housekeeping' gene set, their selection is based on biology as well as on their experimental calculation across thousands of observations under numerous experimental conditions. This selection of a 'house-keeping' set has been modelled by labelling the same mRNA sample from the same cell line by both Cy3 and Cy5. The ratio distribution estimated by the actual data from the housekeeping set closely predicts the spread of the ratio. Furthermore, the peak of the ratio distribution is correctly calibrated to 1, which is what we expected as the two samples should be essentially identical. The significance of basing our measurement on the analytical ratio distribution is that we can associate a confidence interval to each ratio measurement so that a significant difference can be easily detected (15). More importantly, this approach allows us to associate a p-value to each ratio measurement. Finally, it is possible to derive quality measurements from the above equation.

4.2.6 Multiple image analysis (gene clustering)

The value of microarray technology is not only that it enables a fast screening method for the expression of individual genes, but also in that it enables the investigator to study gene interactions in parallel. One application is to use cDNA microarray gene expression patterns to look for relatedness among cell types (17). While the previous discussion has focused on single-image (slide) analysis reports of genes with significantly different expression level between two probes, multiple images (slides) can be analysed to explore the temporal expression pattern for a given gene, or to study the similarity between expression patterns from different samples (e.g. patterns of expression between stages of cancer progression).

Data clustering is one of the data exploration techniques in which any non-random patterns or structures requiring further explanation are recognized and placed into a small number of homogeneous groups or clusters. Applications where the investigator wishes to view the expression of genes across multiple slides (experiments) are too numerous to be listed here. Accordingly, one of the data mining objectives of our laboratory is to cluster genes based on their statistical behaviour such that some functional relationship may be hypothesized, or the characteristics of expression patterns for all clusters may be extracted for fingerprinting purposes (18).

Use of the multidimensional scaling (MDS) technique for visualization of biological sample similarity (17), a *k*-means-based algorithm and a hierarchical clustering method similar to (18) for gene expression similarity clustering and visualization, along with an efficient fingerprinting gene selection method for each biologically meaningful cluster, is being modelled in our laboratory to analyse data from multiple images. For example, expression array images are obtained at different time points and their ratios are extracted. The gene expression profiles can be either normalized ratio data or the quantified ratio data based on their confidence interval for all time points. Combined with other data-mining techniques, an appropriate number of clusters is identified interactively. An example is shown in *Plate 3(F)* where multidimensional scaling is used and two clusters can be observed visually. On the other hand, the individual genes can be identified based on the similarity of their expression profiles and placed together if they are closely related, as shown in *Plate 3(G)*. The statistical cluster may be further divided based on a known biological/biochemical function or pathway.

4.2.7 Microarray analysis workbench

A set of software tools has been developed within our laboratory at the Cancer Genetics Branch at NHGRI for the analysis of microarray data. The analysis workbench is a collection of IPLab extensions for the Macintosh computer (IPLab is an image processing package by Scanalytics). Some of tools include: (i) LoadSKN, which loads images from the NIH scanner or other scanning instruments; (ii) AlignArray, which aligns two images in case the images from red and green channels were scanned separately; (iii) DeArray, which is the central processing tool developed by our laboratory for controlling most of image processing tasks, including target segmentation, background intensity estimation and probe intensity extraction; and (iv) TargetLocator, which reports target information, refines statistics, and performs some image enhancement tasks. The development of the microarray analysis workbench software is a continuous process, linked with the perfection of microarray technology and the progress of various applications of the technology.

5 Analysis of large-scale expression data

Microarray technologies are providing the means with which to perform large-scale and whole-genome expression studies. In addition to pushing the envelope on our current understanding of gene expression and regulation, it will also present substantial challenges in the areas of data management and analysis. For this reason, it becomes important to focus on effective informatics methods in order to make novel biological conclusions.

To this end, we have developed ArrayDB 2.0, a software suite that provides an interactive user interface for the mining and analysis of microarray gene expression data. All of the analysed data from a microarray experiment, as well as information about the clones and experimental conditions used in the experiment, are stored in a relational database (19). Newly developed upload tools allow individual investigators to directly populate the database through a Web front-end. The data itself can then be viewed by the user through the use of CGI scripts and Java applets. The CGI scripts provide connections to the relational database, and the applets provide a graphical representation of the experimental data. The two main applets are the Experiment Viewer and the Multi-Experiment Viewer. The Experiment Viewer presents the expression data for a single experiment in histogram form, allowing specific areas of the histogram to be selected. Once a selection is made, the actual microarray slide image is displayed, as well as detailed information about each of the clones in the selected range. Each clone is also linked to a variety of external databases, such as UniGene, dbEST, and GeneCards, whenever such information is available. Most importantly, links are provided to the Kyoto Encyclopedia of Genes and Genomes (KEGG), showing relevant biological pathways and putting the experiment in metabolic context. The Multi-Experiment Viewer allows the user to query a range of related experiments at one time in order to see changes in the pattern of expression of the same gene over different experiments.

The software is currently and freely available for Sybase/UNIX and is in the process of being ported to Oracle/UNIX. Detailed information about the program itself can be found on the ArrayDB Web site (http://genome.nhgri.nih.gov/arraydb).

6 Future directions

Microarray technology is still a new yet immensely powerful tool in molecular biology. The demand for parallel gene expression analysis by cDNA microarray and the dramatic increase in the number of laboratories around the world which have implemented, established, and improved the technology will lead to its rapid evolution. Improvements are anticipated for every aspect of microarray technology. The quantity and quality of sequence verified clones available to be printed is improving at a rapid rate. The production of microarray DNA chips containing all the genes in the human genome requires only about an order of magnitude increase in density of spots on glass slides and the forthcoming sequencing of the human genome. Efforts to improve labelling methodologies are aiming to decrease the amount of RNA required for hybridizing reactions allowing analysis of biological samples where tissue is limited. To this end, signal amplification based indirect labelling approaches appear promising and other approaches to reducing the input mRNA requirements are required. Already, high sensitivity and high throughput expression profiling has in many ways revolutionized gene expression experimentation and analysis. New bioinformatic approaches are currently being developed to better mine, manage, and analyse large-scale expression data and maximize the amount of biologically relevant information that can be extracted. Differential gene expression experiments have already been successful in a wide spectrum of application to identify single genes and group of genes that are important in various biological processes. In the future, however, analysis of data across thousands of experiments will enable the elucidation of gene expression fingerprints that can be associated with specific physiological or pathological states. It is becoming clear that we have only scratched the surface of what can be accomplished with the creative application of microarray technology.

Acknowledgements

The custom-built robotic arrayer was developed by Drs Stephen B. Leighton and scanner optics by Paul D. Smith. Thomas Pohida developed the electronics of both the arrayer and scanner. We thank Yuan Jiang, Gerald C. Gooden, John Lueders, Kim A. Gayton, Art A. Glatfelter, and Robert L. Walker for their excellent technical assistance on this work. A host of talented investigators have also contributed to the NIH Microarray Project, including: J. Kahn, D, Duggan, M. Boguski, G. Schuler, O. Ermolaeva, T. Pohida, J. Hudson, A. Fornace, S. Amundson, S. Zeichner, C. Xiang, R. Simon, J. DeRisi, and P. Brown. We also thank Darryl Leja for assistance in the preparation of the illustrations.

References

1. Collins, F. S., Patrinos, A., Jordan, E., Chakravarti, A., Gesteland, R., and Walters, L. (1998). *Science*, **282** (5389), 682–9.
2. Schena, M., Shalon, D., Davis, R. W., and Brown, P. O. (1995). *Science*, **270** (5235), 467–70.
3. DeRisi, J., Penland, L., Brown, P. O., Bittner, M. L., Meltzer, P. S., Ray, M., Chen, Y., Su, Y. A., and Trent, J. M. (1996). *Nat. Genet.*, **14** (4), 457–60.
4. Lockhart, D. J., Dong, H., Byrne, M. C., Follettie, M. T., Gallo, M. V., Chee, M. S., Mittmann, M., Wang C., Kobayashi, M., Horton, H., and Brown, E. L. (1996). *Nat. Biotechnol.*, **14**, 1675–80.
5. Velculescu, V. E., Zhang, L., Vogelstein, B., and Kinzler, K. W. (1995). *Science*, **270** (5235), 484–7.
6. Liang, P. and Pardee, A. B. (1992). *Science*, **257** (5072), 967–71.
7. Adams, M. D., Kelley, J. M., Gocayne, J. D., Dubnick, M., Polymeropoulos, M. H., Xiao, H., Merril, C. R., Wu, A., Olde, B., Moreno, R. F., *et al.* (1991). *Science*, **252** (5013), 1651–6.
8. Hubank, M. and Schatz, D. G. (1994). *Nucleic Acids Res.*, **22** (25), 5640–8.
9. Iyer, V. R., Eisen, M. B., Ross, D. T., Schuler, G., Moore, T., Lee, J. C.F, Trent, J. M., Staudt, L. M., Hudson, J. Jr, Boguski, M. S., Lashkari, D., Shalon, D.,Botstein, D., and Brown, P. O. (1999). *Science*, **283** (5398), 83–7.
10. Adams, M. D. *et al.* (1991). *Science*, **252**, 1651–6.
11. Benson, D. A., Boguski, M. S., Lipman, D. J., and Ostell, J. (1997). *Nucleic Acids Res.*, **25**, 1–6.
12. Trent, J. M., Bittner, M., Zhang, J., Wiltshire, R., Ray, M., Su, Y., Gracia, E., Meltzer, P., De Risi, J., Penland, L., and Brown, P. (1997). *Clin. Exp. Immunol.*, **107** (Suppl. 1), 33–40.
13. Bowtell, D. D. (1999). *Nat. Genet.*, **21** (Suppl. 1), 25–32.
14. Chen, J. J., Wu, R., Yang, P. C., Huang, J. Y., Sher, Y. P., Han, M. H., Kao, W. C., Lee, P. J., Chiu, T. F., Chang, F., Chu, Y. W., Wu, C. W., and Peck, K. (1998). *Genomics*, **51** (3), 313–24.

15. Chen, Y., Dougherty, E., and Bittner, M. (1997). *J. Biomed. Optics*, **2**, 364.
16. Wang, X. F. and Herman, B. (ed.) (1996). *Fluorescence imaging spectroscopy and microscopy*. John Wiley & Sons, Inc., NY.
17. Khan, J. *et al.* (1998). *Cancer Res.*, **58**, 5009–13.
18. Eisen, MB., Spellman, PT., Brown, P. O., and Botstein, D. (1998). *Proc. Natl Acad. Sci. USA*,. **95**, 14863–8.
19. Ermolaeva, O., Rastogi, M., Pruitt, K. D., Schuler, G. D., Bittner, M. L., Chen, Y., Simon, R., Meltzer, P., and Trent, J. M. (1998). *Nat. Genet.*, **20**, 19–23.

Chapter 7

SADE: a microassay for serial analysis of gene expression

Lydie Cheval, Bérangère Virlon, Jean-Marc Elalouf
Département de Biologie Cellulaire et Moléculaire, Service de Biologic Cellulaire, CNRS URA1859, CEA Saclay, g11g1 Gif-sur-Yvette Cedex, France

1 Introduction and overview

1.1 High-throughput methods for quantitative analysis of gene expression

Several methods are now available for monitoring gene expression on a genomic scale. These include DNA microarrays (1, 2) and macroarrays (3, 4), expressed sequence tag (EST) determination (5, 6), and the serial analysis of gene expression (SAGE) method (7). Such methods have been designed, and are still used, for analysing macroamounts of biological material (1–5 μg of poly(A) mRNAs, i.e. $\sim 10^7$ cells). However, mammalian tissues consist of several different cell types with specific physiological functions and gene expression patterns. Obviously, this makes intricate the interpretation of large-scale expression data in higher organisms. It is therefore most desirable to set out methods suitable for the analysis of defined cell populations.

SAGE has been shown to provide rapid and detailed information on transcript abundance and diversity (7–10). It involves several steps for mRNA purification, cDNA tags generation and isolation, and polymerase chain reaction (PCR) amplification. We reasoned that increasing the yield of the various extraction procedures, together with slight modifications in the number of PCR cycles could enlarge SAGE potentiality. Here we present a microadaptation of SAGE, referred to as SADE (11) as, in contrast to the original method, it allows to provide quantitative gene expression data on a small number (30 000–50 000) of cells.

1.2 Protocol outline

SAGE was first described by Velculescu *et al.* in 1995 (7), and rests on three principles which have now been all corroborated experimentally: (i) short nucleotide sequence tags (10 bp) are long enough to be specific of a transcript, especially if they are isolated from a defined portion of each transcript; (ii) concatenation of several tags within a single DNA molecule greatly increases the throughput of data acquisition; (iii) the quantitative recovery of transcript-specific tags allows to establish representative gene expression profiles.

Figure 1, modified from the original studies of Velculescu *et al.*, summarizes the different steps of SADE, a Sage Adaptation for Downsized Extracts (11b). Briefly, mRNAs are extracted using oligo(dT)$_{25}$ covalently bound to paramagnetic beads. Double strand (ds) cDNA is synthesized from mRNA using oligo(dT)$_{25}$ as primer for the first strand synthesis. The cDNA is then cleaved using a restriction endonuclease (anchoring enzyme: SAu3AI) with a 4-bp recognition site. As such an enzyme cleaves DNA molecules every 256 bp (4^4) on average, virtually all cDNAs are predicted to be cleaved at least once. The 3′ end of each cDNA is isolated using the property of the paramagnetic beads and divided in half. Each of the two aliquots is ligated via the anchoring enzyme restriction site to one of the two linkers containing a type IIS recognition site (tagging enzyme: *Bsm*FI) and a priming site for PCR amplification. Type IIS restriction endonucleases display recognition and cleavage sites separated by a defined length (14 bp for *Bsm*FI), irrespective of the intercalated sequence. Digestion with the type IIs restriction enzyme thus releases linkers with an anchored short piece of cDNA, corresponding to a transcript-specific tag. After blunt ending of tags, the two aliquots are linked together and amplified by PCR. As all targets are of the same length (110 bp) and are amplified with the same primers, potential distortions introduced by PCR are greatly reduced. Furthermore, these distortions can be evaluated, and the data corrected accordingly (7, 8). Ditags present in the PCR products are recovered through digestion with the anchoring enzyme and gel purification, then concatenated and cloned.

1.3 Differences between SAGE and SADE

In the SAGE method, mRNAs are isolated using conventional methods, then hybridized to biotinylated oligo(dT) for cDNA synthesis. After cleavage with the anchoring enzyme, the biotinylated cDNA fraction (3′ end) is purified by binding to streptavidin beads. In the SADE method, mRNAs are directly isolated from the tissue lysate through hybridization to oligo(dT) covalently bound to magnetic beads. Then, all steps of the experiment (until step 3 of *Protocol 5*) are performed on magnetic beads. This procedure saves time for the initial part of the experiment and, more importantly, provides better recovery. Quantitative analysis of the cDNA amounts available for library construction revealed dramatic differences between SAGE and SADE. With the SAGE method, starting from 500 mg tissue we obtained 1.7 μg of cDNA, and only 4 ng were able to bind to streptavidin beads after *Sau*3AI digestion. With the SADE method, starting from 250 mg of tissue, 3.2 μg of cDNA were synthesized on beads, and 0.5 μg remained bound after *Sau*3AI cleavage. The increased yield of SADE (×250) explains our success in constructing libraries from as few as 30 000 cells. Using either biotinylated oligo(dT) designed by us or obtained from mRNA purification kits, we always get poor cDNA recoveries. This may be explained by the fact that the binding of

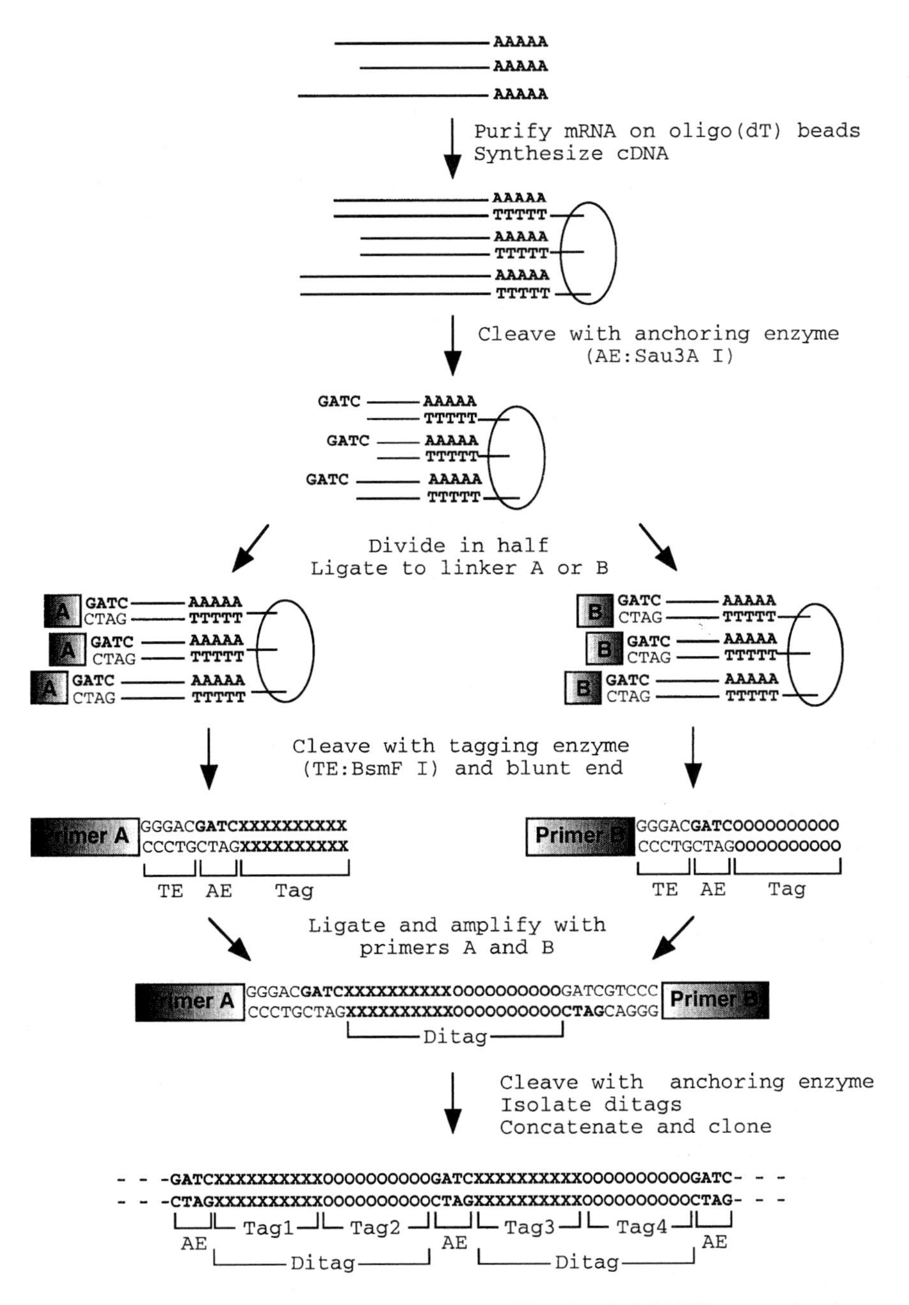

Figure 1 Outline of procedures for constructing SADE libraries. Poly(A) RNAs are isolated from tissue lysate using oligo$(dT)_{25}$ covalently linked to paramagnetic beads, and cDNA is synthesized under solid-phase condition. Bold face characters correspond to biologically relevant sequences, whereas light characters represent linker-derived sequences. The anchoring enzyme (AE) is *Sau*3AI, whereas the tagging enzyme (TE) is *Bsm*FI. See text for details.

biotinylated DNA can be altered by several parameters, such as the amount of free biotin and the presence of phenol, the length and composition of the biotinylated DNA fragments, and the length of the spacer between the oligo(dT) and the biotin molecule.

Another important difference between SAGE and SADE concerns the selected anchoring enzyme. Although any restriction enzyme with a 4-bp restriction site could serve as an anchoring enzyme, *Sau*3AI was preferred to *Nla*III (7–10) or other enzymes in our studies. Several cDNA libraries used for large-scale sequencing are constructed by vector priming, followed by cDNA cleavage with *Mbo*I (an isoschizomer of *Sau*3AI which does not cut the vector (methylated) DNA), and circularization (6). SADE tags therefore correspond to the cDNAs 5′ end of these libraries, which enables to use more efficiently EST databases to analyse our data.

2 Tissue sampling and mRNA isolation

2.1 Tissue sampling and lysis

The initial steps of library construction require the usual precautions recommended for experiments carried out with RNAs (12). In addition, as library construction involves large-scale PCR (*Protocol* 6), care must be taken to avoid contamination from previous libraries. Working under PCR grade conditions is especially important when low amounts of tissue or cells are used.

Starting from whole tissues (i.e. kidney, liver, brain, etc.), we routinely use the following procedures. After animal anaesthesia or decapitation, the tissue is removed as quickly as possible, rapidly rinsed in ice-cold phosphate-buffered saline, sliced in ~50 mg pieces, and frozen in liquid nitrogen. The frozen sample is then ground to a fine powder under liquid nitrogen using a mortar and a pestle, transferred into lysis binding buffer (*Protocol 1*), and homogenized with a Dounce tissue disrupter. To avoid loss of material, small samples (≤20 mg) can be transferred without previous freezing in the lysis binding buffer, and homogenized in a 1 ml Dounce. The respective amounts of tissue and lysis binding buffer needed for a variety of conditions are indicated in *Table 1*.

Starting from isolated or cultured cells, the procedure is much more rapid. The cell suspension, maintained in appropriate culture or survival medium, just

Table 1 Small and large scale mRNA isolation and cDNA synthesis

Tissue/cells	**Reaction volume (μl)** **Lysis binding buffer (ml)**	**Oligo(dT) beads (μl)**	**1st strand (μl)**	**2nd strand (μl)**
250 mg/3 × 10^7	5.50	600	50	400
30 mg/3 × 10^6–6 × 10^6	0.70	100	50	400
4 mg/10^5–10^6	0.10	30	25	200
0.5 mg/3 × 10^4–10^5	0.05–0.10	20	25	200

needs to be centrifuged at 600–1200 *g* for 5 min. After supernatant removal, the lysis binding buffer is added on to the cell pellet, and the sample is homogenized by vortexing. This procedure has been successfully applied to $3 \times 10^4 - 3 \times 10^7$ cells (*Table 1*).

2.2 mRNA isolation

Protocols 1–7 describe the generation of a SADE library from 0.5 mg of tissue. The amount of cDNA recovered corresponds to an experiment carried out on the mouse kidney. Slightly different amounts are expected to be obtained from other tissues, according to their mRNA content. The procedures described herein have been repeatedly used without modifications with $3 \times 10^4 - 10^5$ isolated cells. As some applications can be performed on large amounts of tissue or cells, protocol adaptations and anticipated results for these kinds of experiments are also provided.

In our initial experiments, RNAs were extracted using standard methods (13), and poly(A) RNAs were isolated on oligo(dT) columns. Besides being time consuming, this procedure provides low and variable mRNA amounts, and cannot be easily scaled down. The alternative procedure described here (use of oligo$(dT)_{25}$ covalently linked to paramagnetic beads) is a single tube assay for mRNA isolation from tissue lysate. In our hands, it yields four times higher mRNA amounts than standard methods. Kits and helpful instructions for mRNA isolation with oligo(dT) beads can be obtained from Dynal. Handling of these beads is relatively simple, but care must be taken to avoid centrifugation, drying or freezing, as all three processes are expected to lower their binding capacity. On the other hand, beads can be resuspended by gentle vortexing or pipetting without extreme precautions.

Protocol 1

mRNA purification

Equipment and reagents

- Appropriate tissue or cells
- Dynabead mRNA direct kit (Dynal, ref. no. 610-11) containing Dynabeads oligo$(dT)_{25}$, lysis binding buffer, and washing buffers
- Magnetic Particle Concentrator (MPC) for 1.5 ml tubes (Dynal, ref. no. 12004)
- 5 × reverse transcription (RT) buffer (250 mM Tris–HCl (pH 8.3), 375 mM KCl, 15 mM $MgCl_2$), provided with cDNA synthesis kit (see Protocol 2)
- Glycogen for molecular biology (Boehringer Mannheim, ref. no. 901393)

Method

1 Lyse the tissue sample in 100 µl lysis binding buffer supplemented with 10 µg glycogen.

2 Add 20 µl of Dynabeads in a 1.5 ml tube and condition them according to manufacturer's instructions.

Protocol 1 continued

3. Using the MPC, remove the supernatant from the Dynabeads and add the tissue lysate (100 μl). Mix by vortexing and anneal mRNAs to the beads by incubating 10 min at room temperature.
4. Place the tube 2–5 min in the MPC and remove the supernatant. The mRNAs are fixed on the beads.
5. Using the MPC, perform the following washes (all buffers contain 20 $\mu g\,ml^{-1}$ glycogen): twice with 200 μl washing buffer containing lithium dodecyl sulphate (LiDS), three times with 200 μl washing buffer, and twice with 200 μl ice-cold 1 × RT buffer. Resuspend the beads by pipetting, transfer the suspension in a fresh 1.5 ml tube, wash once with 200 μl ice-cold 1 × RT buffer and immediately proceed to Protocol 2. mRNAs on the beads are now ready for first strand cDNA synthesis.

2.3 mRNA integrity and purity

Before generating a cDNA library, it is generally advised to check for mRNA integrity by Northern blot analysis. However, this control experiment consumes part of the material, takes several days, and often leads to ambiguous results (a variety of reasons can cause poor Northern hybridization signals). In addition, it is no longer possible when using small amounts of tissue or cells. RNA degradation has only to be expected in the three following conditions: (i) cell survival is not maintained before lysis or freezing; (ii) cell thawing outside of lysis buffer; and (iii) use of poor quality reagents. As RNase-free reagents are now available from a variety of company, it is much more rapid and effective to check for survival (i.e. select the appropriate culture medium) and freezing conditions than to perform tricky tests on RNA aliquots.

The purity of mRNAs isolated with oligo(dT) beads is better than that obtained with conventional methods. When we generated SAGE libraries using mRNAs extracted with guanidinium thiocyanate and oligo(dT) columns, nuclear encoded rRNAs amounted to 1% of the sequenced tags. Using the alternative mRNA extraction procedure, rRNAs tags are no longer present in the library.

3 First and second strand synthesis

3.1 First strand cDNA synthesis

The first step in the synthesis of cDNA is copying the mRNA template into complementary single-strand cDNA. In *Protocol 2*, first strand cDNA is synthesized using Moloney Murine Leukaemia Virus RT (MMLV RT). With this enzyme, we have been able to generate SADE libraries from either large or minute amounts of cells (*Table 1*). In our last series of experiments, we have however used SuperScript II MMLV RT, provided with the SuperScript cDNA synthesis kit (Life Technologies, ref. no. 18090-019). In this case, the amount of cDNA formed (see Section 3.3) was increased approximately fourfold. Although this better yield likely results from both the synthesis of longer cDNAs (which is not essential for the current application) and of a higher number of cDNA molecules, we strongly recommend the use of SuperScript II MMLV RT for very small samples (≤50 000 cells). The protocol will be similar to the one described here, except for reaction volumes (20 μl for first strand synthesis, and 150 μl for second strand synthesis).

mRNAs are generally heated 5 min at 65 °C before RT to break up secondary structures. As such a high temperature will also denature the mRNA-oligo$(dT)_{25}$ hybrid, we only heat the sample at 42 °C before initiation of the first strand synthesis.

3.2 Second strand synthesis

Many procedures have been developed for second strand cDNA synthesis. The method used here is a modification of the Grubler and Hoffman procedure. Briefly, the mRNA (in the mRNA-cDNA hybrid) is nicked by *Escherichia coli* RNAse H. *E. coli* DNA polymerase initiates the second strand synthesis by nick translation. *E. coli* DNA ligase seals any breaks left in the second strand cDNA. The procedure is described in *Protocol 2*. This step is usually very efficient (approximately 100%) so that a 2 h-incubation period is sufficient when starting from macroamounts of material (>100 mg of tissue or 10^7 cells).

Protocol 2

cDNA synthesis and cleavage

Equipment and reagents

- cDNA synthesis kit (Life Technologies, ref. no. 18267-013) contains all buffers and enzymes necessary for first and second strand cDNA synthesis
- [α-^{32}P]dCTP 6000Ci mmol^{-1} (Amersham, ref. no. AA0075)
- TEN (10 mM Tris-HCl (pH 8.0), 1 mM ethylenediamine tetraacetic acid (EDTA), 1 M NaCl)
- Restriction endonuclease *Sau*3AI 4 U μl^{-1} (New England Biolabs, ref. 169L), provided with 10 × reaction buffer and purified 100 × bovine serum albumin (BSA; 10 mg ml^{-1})
- Magnetic Particle Concentrator (MPC) (Dynal)
- Geiger counter
- Automated thermal cycler or water-baths equilibrated at 42 °C, 37 °C, and 16 °C

Method

1. Resuspend the beads in 12.5 μl of 1 × first strand (i.e. RT) reaction buffer.
2. Incubate 2 min at 42 °C.
3. Place the tube at 37°C for 2 min. Add 12.5 μl of the following mix: 5 μl DEPC-treated water, 2.5 μl 5 × first strand buffer, 1.25 μl dNTP 10 mM, 2.5 μl dithiothreitol (DTT) 100 mM, 1.25 μl MMLV reverse transcriptase.
3. Incubate 1 h at 37 °C and chill on ice.
4. On ice, prepare the following mix: 169.7 μl DEPC-treated water, 4.5 μl dNTP 10 mM,

Protocol 2 continued

24 μl second strand buffer, 2 μl [α-^{32}P]dCTP, 6 μl *E. coli* DNA polymerase I, 1.05 μl *E. coli* RNAse H, 0.75 μl *E. coli* DNA ligase, 2 μl glycogen 5 μg μl^{-1}.

5. Add 175 μl to the first strand tube and incubate overnight at 16 °C. Keep the remaining mix for subsequent measurement of its radioactivity and calculation of dCTP specific activity.
6. Wash beads to remove non incorporated [α-^{32}P]dCTP: four times with 200 μl TEN + BSA,[a] and three times with 200 μl ice-cold 1 × mix *Sau*3AI + BSA.[a] Check with Geiger counter that the last eluate is not radioactive, whereas the material bound on the beads is highly radioactive.
7. Add on the beads the following mix: 88 μl H_2O, 10 μl 10 × mix *Sau*3AI, 1 μl 100 × BSA, 1 μl *Sau*3AI. Incubate 2 h at 37 °C. Vortex intermittently.
8. Chill 5 min on ice.
9. Using the MPC, remove the supernatant, which contains the 5′ end of the cDNA. Wash once with 200 μl of 1 × mix *Sau*3AI + BSA[a]. Remove this second supernatant, pool it with the first one, and store the resulting solution (300 μl) in order to measure the yield of second strand synthesis (see text). Before going to step 10, check with Geiger counter that both the eluate and beads are radioactive.
10. Resuspend the beads in 200 μl TEN supplemented with BSA.[a]

[a] Final concentration of BSA: 0.1 mg ml^{-1}.

3.3 Yield of second strand synthesis

A method to calculate the yield for first and second strand cDNA synthesis is given in the cDNA synthesis kit instruction manual. We do not measure the yield of first strand cDNA synthesis because, as discussed above (Section 2.3), this implies to set away part of the preparation.

The amount of ds cDNA formed is calculated by measuring radioactivity incorporation in the 5′ end of the cDNA, which is released in the supernatant after *Sau*3AI digestion (see *Protocol 2*). The 300 μl supernatant is extracted with phenol–chloroform–isoamyl alcohol (PCI) and the ds cDNA is ethanol precipitated in the presence of glycogen (50 μg ml^{-1}) and 2.5 M ammonium acetate. The pellet is resuspended in 8 μl of TE. Half of the material is used for liquid scintillation or cerenkov counting, and the remaining is loaded on a 1.0 or 1.5% agarose gel. For experiments carried out on 250, 30, 4, and 0.5 mg of mouse kidney, we obtained the following amounts (μg) of ds cDNA: 2.8, 0.3, 0.05, and 0.01. The higher amount corresponds to the incorporation of 1.3% of the input radioactivity. In these experiments, three of the four cDNA samples could be detected by ethidium bromide staining after gel electrophoresis (*Figure 2*). Their size ranged between $<$0.2 and ~3 kbp (the small size of most cDNA fragments is due to *Sau*3AI digestion). When cDNA amounts are below the detection threshold of the ethidium bromide staining method, autoradiographic analysis can be per-

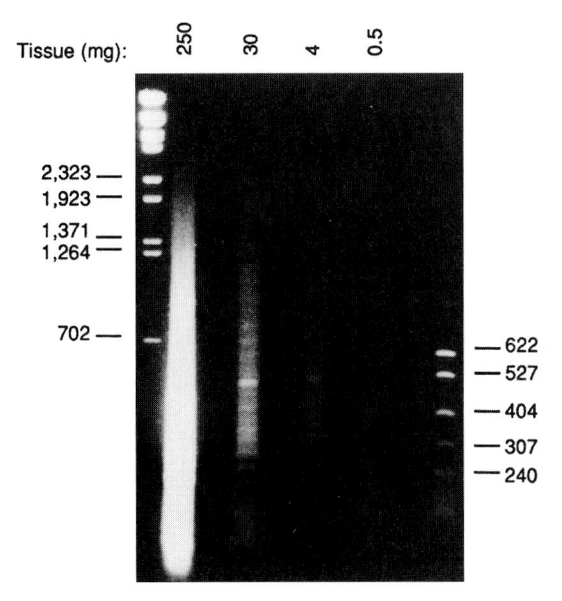

Figure 2 Gel analysis of cDNAs synthesized from different amounts of tissue. Poly(A) RNAs were isolated from the indicated amounts of mouse kidney, and cDNAs were synthesized and *Sau*3AI-digested on paramagnetic beads (see *Protocols 1* and *2*). cDNAs released from beads were recovered, and half of the material obtained from each reaction was analysed on a 1% agarose gel stained with ethidium bromide. Position of molecular weight markers are indicated in bp: left, λ *Bst*EII-digest; right, pBR *Msp*I-digest.

formed. In this case, the gel is fixed in 10% acetic acid, vacuum dried and exposed overnight at −80 °C with one intensifying screen for autoradiography.

4 Linkers design, preparation, and ligation

4.1 Linkers design

A variety of linkers can be used at this point. Linkers must contain three important sequences: (i) the appropriate anchoring enzyme overhang; (ii) a recognition site for a type IIs restriction enzyme (tagging enzyme); (iii) a priming site for PCR amplification. High-quality linkers are crucial for successful library generation.

Table 2 provides the sequence of linkers and PCR primers used in our experiments. All four linkers must be obtained gel-purified. Linkers 1B and 2B display two modifications: (i) 5′ end phosphorylation, and (ii) C7 amino modification on the 3′ end. Linkers phosphorylation can be performed either enzymatically with T4 polynucleotide kinase, or chemically at the time of oligonucleotide synthesis. In both cases, phosphorylation efficiency must be tested (*Protocol 3*). We use chemically phosphorylated linkers. Linkers modification on the 3′ end serves to increase the efficiency of ditag formation (*Protocol 5*, steps 8–11). Indeed, the modified 3′ end cannot be blunt-ended and will not ligate to cDNA tags or linkers.

Table 2 Sequence of linkers and PCR primers

Oligonucleotide	Sequence
Linker 1A	5′-TTTTGCCAGGTCACTCAAGTCGGTCATTCATGTCAGCACAGGGAC-3′
Linker 1B[a]	5′-GATCGTCCCTGTGCTGACATGAATGACCGACTTGAGTGACCTGGCA-3′
Linker 2A	5′-TTTTTGCTCAGGCTCAAGGCTCGTCTAATCACAGTCGGAAGGGAC-3′
Linker 2B[a]	5′-GATCGTCCCTTCCGACTGTGATTAGACGAGCCTTGAGCCTGAGCAA-3′
Primer 1	5′-GCCAGGTCACTCAAGTCGGTCATT-3′
Primer 2	5′-TGCTCAGGCTCAAGGCTCGTCTA-3′

[a] Linkers 1B and 2B include two modifications (5′-phosphorylation and 3′-C7 amino modification).

With regard to the PCR priming site, it was designed with the help of Oligo™ software (Medprobe, Norway) in order to obtain PCR primers with high Tm (60 °C), and avoid self-priming or sense/anti-sense dimer formation. Two different priming sites must be designed in 'left' and 'right' linkers, otherwise the target will undergo panhandle formation, and thus escape PCR amplification.

Protocol 3

Preparing and testing linkers

Equipment and reagents

- Linkers 1A, 1B, 2A, and 2B at 20 pmol μl^{-1}
- Primers 1 and 2 at 20 pmol μl^{-1}
- T4 DNA ligase 1 U μl^{-1} (Life Technologies, ref. 15224-017) and 5 × reaction buffer
- 10 mM adenosine triphosphate (ATP)
- PCR reagents: Taq DNA polymerase 5 U μl^{-1} (Eurobio), 10 × PCR buffer (200 mM Tris-HCl (pH 8.3), 15 mM $MgCl_2$, 500 mM KCl, 1 mg ml^{-1} gelatin), 1.25 mM dNTP, 100 mM $MgCl_2$, and 100 mM DTT
- Tris-HCl buffered (pH 7.9) PCI
- Restriction endonucleases *Sau*3AI (4 U μl^{-1}) and *Bsm*FI (2 U μl^{-1}) (New England Biolabs, refs. 169L and 572L), provided with 10 × reaction buffer and 100 × BSA
- 10-bp DNA ladder (Life Technologies, ref. 10821-015)
- Automated thermal cycler and water baths equilibrated at 14 °C, 37 °C, and 65 °C
- 10 M ammonium acetate
- Tris-EDTA (TE) buffer (10 mM Tris-HCl (pH8.0), EDTA 1 mM)

Method

1. Mix 25 µl of linker 1A and 25 µl of linker 1B in a 0.5 ml PCR tube (final concentration: 10 pmol μl^{-1}). Proceed similarly for linkers 2A and 2B.
2. Transfer PCR tubes in the thermal cycler. Heat at 95 °C for 2 min, then let cool at room temperature for 20 min on the bench. Store at −20 °C.
3. Test self-ligation of each hybrid , as well as ligation of hybrid (1A/1B) with hybrid (2A/2B). Set up three ligation reactions by mixing 1 µl of hybrid (1A/1B) (tube 1), 1 µl of hybrid (2A/2B) (tube 2), 0.5 µl of hybrid (1A/1B) and 0.5 µl of hybrid (2A/2B) (tube 3) with 2 µl 10 mM ATP, 4 µl 5 × ligase mix, 12 µl H_2O, and 1 µl T4 DNA ligase.

Protocol 3 continued

4. Incubate 2 h or overnight at 14 °C. Analyse 10-μl aliquots on a 3% agarose gel using 10-bp DNA ladder as marker. Most of the material (⩾80%) consists of a 94-bp DNA fragment.
5. Proceed to PCR using 10^5 targets from tube 3 reaction (dilute using TE buffer supplemented with 0.1 mg ml^{-1} BSA). Mix 1 μl of diluted ligation product, 5 μl 10 × PCR buffer, 1 μl 100 mM $MgCl_2$, 8 μl 1.25 mM dNTP, 2.5 μl primer 1, 2.5 μl primer 2, and 30 μl water. Prepare four such reactions and a control tube without linker, transfer in the thermal cycler and heat at 80 °C for 2 min.
6. Add in each tube 50 μl of Taq polymerase amplification mix (5 μl 10 × PCR buffer, 4 μl 100 mM DTT, 0.5 μl Taq polymerase, 40.5 μl water), and 60 μl of mineral oil if necessary for your thermal reactor.
7. Perform 29 PCR cycles (95 °C, 30 s; 58 °C, 30 s; 70 °C, 45 s), followed by an additional cycle with a 5-min elongation time.
8. Analyse 10-μl aliquots on a 3% agarose gel. A 90-bp amplification fragment is clearly visible.
9. Pool all four PCR samples, extract with an equal volume of PCI. Transfer the aqueous (upper) phase in a fresh tube, then add 100 μl 10 M ammonium acetate and 500 μl isopropanol. Round the tubes several times for mixing, centrifuge (15 000 g) at 4 °C for 20 min, wash twice with 400 μl 75% ethanol, vacuum dry, and resuspend the pellet in 12 μl TE buffer.
10. Set up two 50-μl digestion reactions using 5 μl of DNA and 4 U of *Sau*3AI or *Bsm*FI. Incubate 1 h at 37 °C or 65 °C, as appropriate.
11. Analyse 10-μl aliquots on a 3% agarose gel. Run in parallel 1 μl of uncut PCR product. *Sau*3AI and *Bsm*FI digestion must be completed to ⩾80%.

4.2 Linkers preparation

It is essential to check that ds linkers can be ligated, PCR amplified, and digested with the anchoring and tagging enzyme. Success with the *Protocol 3* experiments is a prerequisite before attempting to prepare a library. The PCR conditions described here have been optimized for Hybaid thermal reactors (TR1 and Touch Down) working under control or simulated tube conditions. Different conditions may be used with other machines. Note that as the target is quite small (90 bp), elongation is performed at a relatively low temperature.

4.3 Ligating linkers to cDNA

The concentration of ds linkers should be adapted to the amount of cDNA used to prepare the library. In the original protocol of Velculescu *et al.*, 2 μg (74 pmol) of ds linkers are used. Considering that starting from 5 μg of mRNAs, 2 μg of cDNAs with 1–2 kb average size are obtained, the amount of cDNA available for ligation is in the range of 1.5–3 pmol. As a large excess of linkers decreases the

PCR signal to noise ratio, we perform ligation with 8 pmol ds linkers for libraries generated from 250 mg of tissue (~5 μg mRNAs). Starting from $5 \times 10^4 - 10^5$ cells (10–40 ng mRNAs), 0.5 pmol of ds linkers are used. A lower amount of linkers may allow efficient ligation, but we have no experience of it.

Protocol 4

Ligating ds linkers to cDNA

Equipment and reagents

- Hybrid (1A/1B) and (2A/2B) at 0.5 pmol μl^{-1}, obtained from Protocol 3, steps 1–2
- TEN, TE, and Low TE (LoTE) buffer (3 mM Tris–HCl (pH 7.5), 0.2 mM EDTA), stored at 4 °C
- 10 × New England Biolabs (NEB) buffer IV reaction buffer and 100 × BSA
- T4 DNA ligase 5 U μl^{-1} (Life Technologies, ref. 15224–041) and 5 × ligation mix; 10 mM ATP
- MPC (Dynal)
- Water-baths equilibrated at 45 °C and 16 °C
- Geiger counter

Method

1. Once the experiments described in Protocol 2 have been carried out, perform two additional washes of the beads before ligating ds linkers to the cDNA. Using the MPC, wash the beads with 200 μl of TEN + BSA.[a] Resuspend the beads in 200 μl of the same buffer (take care to recover the beads completely: mix by repeated pipetting and scrape the tube wall with the pipette tip), then separate into two 100 μl aliquots: one will be ligated to hybrid (1A/1B), the other will be ligated to hybrid (2A/2B).
2. Add 10 μl of fresh Dynabeads in two 1.5 ml tubes. These tubes will be now treated as the two others and will be used as the negative control.
3. Wash twice the four tubes with 200 μl of ice-cold TE buffer + BSA.[a]
4. Immediately after the last rinsing, add to each tube 34 μl of the appropriate mix containing 8 μl of 5 × ligase buffer and 0.5 pmol of hybrid (1A/1B) or hybrid (2A/2B). Heat 5 min at 45 °C then chill on ice.
5. Add in each tube 4 μl of 10 mM ATP, and 2 μl of T4 DNA ligase (final volume: 40 μl). Incubate overnight at 14 °C.
6. Wash beads thoroughly (free linkers will poison the PCR amplification) as follows: four times with 200 μl of TEN + BSA[a] and three times with 200 μl of 1 × NEB IV + BSA.[a] After the first rinsing with NEB IV, take care to resuspend completely the beads (see above) and transfer them to fresh tubes. After the last rinsing, check that radioactivity is still present on the beads, but absent from the supernatant.
7. Proceed to Protocol 5 or store at 4 °C.

[a] Final concentration of BSA: 0.1 mg ml^{-1}.

After ligation (step 5 in *Protocol* 4), it is very important to wash the beads extensively in order to remove free ds linkers. In fact, if ds linkers not ligated to cDNA fragments are not thoroughly eliminated from each sample, the library will contain large amounts (up to 25%) of linkers sequences. This will make data acquisition poorly efficient.

5 Ditags formation

5.1 Release of cDNA tags

Digestion with the tagging enzyme (*Bsm*FI) will release only small DNA fragments from oligo(dT) beads. Consequently, much of the radioactivity remains bound to the beads at this stage. In order to check that extensive rinsing did not cause great loss of material, we usually measure beads radioactivity by Cerenkov counting after *Bsm*FI digestion. For experiments previously described on 250, 30, 4, and 0.5 mg of mouse kidney, the amounts of ds cDNA remaining on the beads reached 450, 67, 13, and 1.8 ng, respectively. Comparison of these data with those dealing with *Sau*3AI-released fragments (Section 3.3) indicates that approximately six times lower cDNA amounts are recovered on beads that on *Sau*3AI supernatants. The average size of *Sau*3AI cut fragments is predicted to be 256 bp. The fraction that remains bound on the beads after *Sau*3AI digestion thus suggests that the average length of cDNA formed is ~1.5 kb, which seems quite reasonable.

The whole amount of *Bsm*FI-released material is used for ditag formation, and we never attempted to quantify it. Nevertheless, the efficiency of *Bsm*FI digestion can be checked when >4 mg of tissue is used for library generation. In this case, a Geiger counter allows the detection of radioactivity in the *Bsm*FI supernatant.

Protocol 5

Release, blunt ending, and ligation of cDNA tags

Equipment and reagents

- *Bsm*FI, 10 × NEB IV buffer, and 100 × BSA
- PCI
- 10 M ammonium acetate
- Sequencing grade T7 DNA polymerase (Pharmacia Biotech, ref. 27098503).
- 5 × mix salt (200 mM Tris–HCl (pH 7.5), 100 mM $MgCl_2$, 250 mM NaCl)
- 3 mM dNTP
- T4 DNA ligase (5 U μl^{-1}) and 5 × reaction buffer
- 100% ethanol, 75% ethanol
- Geiger counter
- Water-baths equilibrated at 65 °C, 42 °C, and 16 °C

Method

1. Remove supernatant and immediately add on the beads 100 μl of the following mix: 87 μl H_2O, 10 μl 10 × NEB IV, 1 μl 100 × BSA, 2 μl *Bsm*FI.
2. Incubate 2 h at 65 °C. Vortex intermittently.
3. Chill 5 min at room temperature, collect the supernatant (which contains the tags)

Protocol 5 continued

and wash beads twice with 75 μl of ice-cold TE + BSA.[a] Pool all three supernatants (250 μl final volume) and add 60 μg glycogen to each of the four reaction tubes. Measure the radioactivity still present on the beads by Cerenkov counting (see text).

4. Add 250 μl (1 volume) PCI to all four supernatants.
5. Vortex, then centrifuge (10 000 g) 10 min at 4 °C. Transfer the upper (aqueous) phase to a fresh tube.
6. Precipitate with high ethanol concentration: add to the aqueous phase 125 μl 10 M ammonium acetate, 1.125 ml 100% ethanol, and centrifuge (15 000 g) 20 min at 4 °C.
7. Wash the pellet twice with 400 μl of 75% ethanol. Vacuum dry and resuspend the pellet in 10 μl LoTE.
8. Add 15 μl of 1 × mix salt on each tube and heat 2 min at 42 °C. Maintain tubes at 42 °C and add 25 μl of the following mix: 3.5 μl H_2O, 5.5 μl 100 mM DTT, 5 μl dNTP mix 1 μl T7 DNA polymerase. Incubate 10 min at 42 °C.
9. Pool together tags ligated to hybrid (1A/1B) and hybrid (2A/2B). Rinse the tubes with 150 μl LoTE + 20 μg glycogen and add this solution to the pooled reactions (final volume: 250 μl). You have now two tubes (one sample, one negative control).
10. Extract with equal volume of PCI and high concentration ethanol precipitate (see steps 4–6). Resuspend the pellet in 6 μl LoTE.
11. Ligate tags to form ditags by adding to the 6 μl sample: 2 μl 5 × mix ligase, 1 μl 10 mM ATP, and 1 μl of T4 DNA ligase (5 U μl^{-1}). Proceed similarly for the negative control, incubate overnight at 16 °C, then add 90 μl LoTE.

[a] Final concentration of BSA: 0.1 mg ml^{-1}.

5.2 Blunt ending of released cDNA tags

Different enzymes may be used for blunt ending *Bsm*FI-released tags. In their original study, Velculescu *et al.* (7) carried out the blunt ending reaction with T4 DNA polymerase. In more recent applications, Klenow DNA polymerase was used (8) and is now recommended. It is also our experience that the success in library generation is very poor using T4 DNA polymerase. This likely comes from the fact that blunt ending with T4 DNA polymerase is carried out at 11 °C (12). Such a low temperature allows protruding termini from unrelated cDNA tags to hybridize, and is thus expected to decrease markedly the amount of material available for the blunt ending reaction. We have successfully used Vent and sequencing grade T7 DNA polymerases to generate blunt ends. The procedure described in *Protocol 5* involves T7 DNA polymerase.

6 PCR amplification

Considering the linkers and primers used in our studies, the desired PCR product is 110-bp long (90 bp of linkers derived sequences, and 20 bp of ditag).

6.1 PCR buffers and procedures

For PCR amplification of ditags, we use buffers and conditions different from those described by Velculescu *et al.*:

(1) Amplification is performed with standard PCR buffers without dimethylsulphoxide and β-mercaptoethanol. Composition of our 10 × PCR buffer is given in *Protocol 3*. Promega buffer (ref. M1901) works equally well. The conditions used in our assay are as follows: 100 μM dNTP, 2.5 mM $MgCl_2$ (2 mM with Promega buffer), 0.5 μM primers, and 5 U Taq polymerase. High amounts of primers and Taq polymerase are used (standard reactions are generally performed with 0.1 μM primers and 1.25 U of enzyme) to ensure a high yield of ditags production. dNTP concentration is also slightly higher (100 vs. 50 μM) than for standard PCR amplifications. Very high dNTP concentrations should nevertheless be avoided as these are known to increase Taq polymerase-dependent misincorporations.

(2) As suggested initially (7), we still perform small-scale PCR, purify the 110-bp fragment, then submit it to preparative PCR.

6.2 Number of PCR cycles

Before starting *Protocol 6*, the optimal number of PCR cycles needs to be determined. This is best accomplished by performing duplicate PCR on 2% of the ligation product and sampling 7-μl aliquots at different cycles. The number of cycles will of course depend on the amount of starting material. For 250 mg tissue pieces, a PCR signal should be obtained with 18 cycles, and the plateau reached at 22–23 cycles. The 110-bp fragment should be largely predominant (amplified products of 90 and 100 bp are not unusual). Examples of PCR carried out on ditags generated from tiny amounts of cells (15 000 to 45 000) are given in *Figure 3*. Using

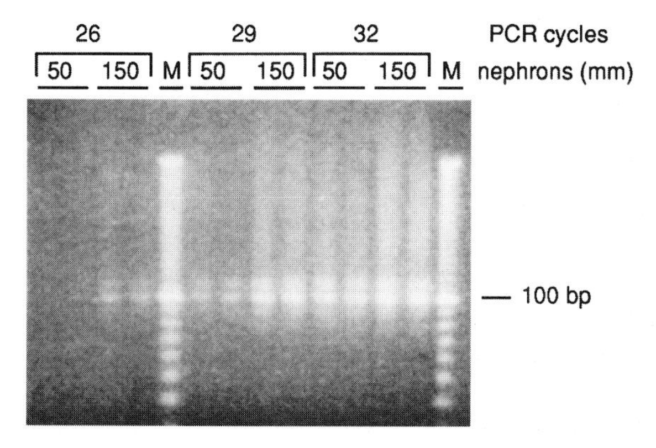

Figure 3 PCR amplification of ditags. Poly(A) RNAs were isolated from 50 or 150 mm of microdissected nephron segments (corresponding to about 15 000 and 45 000 cells, respectively). The corresponding ditags were amplified by PCR using the indicated number of cycles and analysed on a 3% agarose gel stained with ethidium bromide. The expected product (linkers + 1 ditag) is 110-bp long. Molecular weight marker (M) is 10-bp DNA ladder (Life Technologies).

such low amounts of cells, the 110-bp product is no longer predominant. Nevertheless, if maximal yield is achieved with less than 30 cycles (as obtained from 45 000 cells in *Figure 3*), a library which is fairly representative of the tissue can be generated. Small scale PCR (10 reactions, steps 1–5 in *Protocol 6*) is performed on 2 μl and 4 μl aliquots of ligation product for macro- and microamounts of tissue, respectively.

Protocol 6

PCR amplification of ditags

Equipment and reagents

- Automated thermal reactor (Hybaid)
- PCR reagents: Taq polymerase, primers 1 and 2 at 20 pmol μl, 10 × PCR buffer (see Protocol 3), 1.25 mM dNTP, 100 mM $MgCl_2$, and 100 mM DTT
- β-Agarase and 10 × reaction buffer (New England Biolabs, ref. 392L)
- *Sau*3AI reaction buffer and 100 × BSA
- Low melting point (LMP) agarose (Life Technologies, ref. 15517-022)
- TBE 10 × (1.12 M Tris, 1.12 M boric acid, 20 mM EDTA)
- Vertical gel electrophoresis unit, with 20 × 20 cm plates, 1.5 mm thick spacers, and preparative comb
- 10-bp DNA ladder
- Bromophenol blue loading buffer (0.125% bromophenol blue, 10% Ficoll 400, 12.5 mM EDTA) filtered on 0.45-μm membrane

Method

1. Prepare a master mix with the following reagents: 5 μl 10 × PCR Buffer, 1 μl 100 mM $MgCl_2$, 8 μl 1.25 mM dNTP, 2.5 μl primer 1, 2.5 μl primer 2, 27 μl H_2O (multiply these quantities by the number of reactions tubes (usually 12)). Dispense equal aliquots (46 μl) into PCR tubes and add 4 μl of DNA sample (10 tubes), negative control, or H_2O.
2. Transfer the tubes in the thermal reactor and heat 2 min at 80 °C (hot start conditions).
3. Add in each tube 50 μl of the following mix: 5 μl 10 × PCR Buffer, 4 μl 100 mM DTT, 40 μl H_2O, 1 μl Taq Polymerase. Add a drop of mineral oil if necessary according to your thermal cycler.
4. Perform PCR at the following temperatures: 30 s at 94 °C, 30 s at 58 °C, and 45 s at 70 °C (27–30 cycles), followed by one cycle with an elongation time of 5 min.
5. Analyse an aliquot of each tube (7 μl) on a 3% agarose gel using 10-bp DNA ladder as marker.
6. If yield is satisfactory (see Figure 3), pool the 10 PCR tubes in two 1.5 ml tubes. Add 30 μg glycogen in each tube, extract with PCI. Recover the aqueous phase and precipitate the DNA by centrifugation after adding 0.1 volume 3 M sodium acetate and 2.5 volumes 100% ethanol. Wash the pellet with 75% ethanol, vacuum dry, and resuspend in 300 μl TE. Add 75 μl of bromophenol blue loading buffer.

Protocol 6 continued

7 Electrophorese the PCR product through a 3% LMP agarose vertical gel (warm plates 15 min at 55 °C before pouring the gel). Run until bromophenol blue has reached bottom of the gel (~3 h).

8 Cut out the 110-bp fragment from gel and place agarose slice in a 2 ml tube. Add 0.1 volume 10 × β-agarase mix, heat 10 min at 70 °C, then 10 min at 40 °C, and add β-agarase (6 U per 0.2 g of agarose). Incubate 1× h at 40 °C. Add 30 μg glycogen. Extract with PCI and ethanol precipitate as indicated in Step 6. Resuspend the pellet in 300 μl LoTE.

9 After determination of the optimal number of PCR cycles (usually 10), perform large scale PCR (140–150 reactions) using 2 μl of DNA and the Protocol described in Steps 1–4.

10 Pool PCR reactions in 2 ml tubes. Extract with PCI, ethanol precipitate (Step 6), and wash the pellet twice with 75% ethanol. Resuspend the dry pellets in a final volume of 470 μl 1 × mix *Sau*3AI.

6.3 Purification and reamplification

The 110-bp PCR product can be purified either on a 12% polyacrylamide (7) or a 3% agarose slab gel. To avoid overloading and achieve efficient purification, pool no more than 10–12 PCR reactions on an agarose gel and slice agarose as close as possible to the 110-bp fragment. Purification and optimal number of PCR cycles should then be tested on duplicate 2 μl aliquots of the purified product. A single band of 110 bp should now be obtained. The absence of interference from other amplified products is essential to produce large amounts of the 110-bp fragment.

7 Ditags isolation, concatenation, and cloning of concatemers

7.1 Ditags isolation

Two important points need to be addressed for ditag purification. First, as the total mass of linkers is nearly five times that of ditags, a highly resolving polyacrylamide gel is required to purify ditags thoroughly. Second, the short length of ditags makes them difficult to detect on gel by ethidium bromide staining. This problem can be overcome by staining the gel with SYBR Green I (or equivalent products), which ensures a lower detection threshold than ethidium bromide (0.1 instead of 2 ng DNA). To obtain high sensitivity, loading buffer should not contain bromophenol blue (bromophenol blue comigrates with ditags). The gel is stained after migration in a polypropylene or PVC container.

Ditags do not run as a single band on polyacrylamide gel. This may come from subtle effects of base composition on electrophoretic mobility and/or some wobble for *Bsm*FI digestion (7). We cut out from gel all the material ranging from

22 to 26 bp. The elution procedure is labour intensive, but provides ditags that can be concatenated efficiently.

7.2 Concatenation

Starting from 150 PCR reactions, at least 1 μg ditags should be obtained. The optimal ligation time depends on the amount of ditags and on the purity of the preparation. We usually perform ligation for 2 h. When yield is high (⩾1.4 μg), we set up two ligation reactions, and allow them to proceed for 1 or 2 h. The corresponding concatemers are then separately purified on a 8% agarose gel.

Protocol 7

Ditags isolation and concatenation

Equipment and reagents

- *Sau*3AI, 10 × reaction buffer and 100 × BSA
- 50 × TAE (2 M Tris, 57% glacial acetic acid, 50 mM EDTA)
- 12% polyacrylamide gel: 53.6 ml H_2O, 24 ml 40% acrylamide (19:1 acrylamide:bis), 1.6 ml 50 × TAE, 800 μl 10% ammonium persulphate, 69 μl TEMED
- 10-bp DNA ladder
- SYBR Green I stain (FMC Bioproducts, ref. 50513)
- T4 DNA Ligase 5 U μl and 5 × ligation mix; 10 mM ATP
- Vertical gel electrophoresis unit, with 20 × 20 cm plates, 1.5 mm thick spacers, and preparative comb
- Xylene cyanol loading buffer (0.125% xylene cyanol, 10% Ficoll 400, 12.5 mM EDTA)
- SpinX microcentrifuge tubes (Costar, ref. 8160)

Method

1. Save 1 μl of the 110-bp DNA fragment (step 10 of Protocol 6) and digest the remaining by adding 5 μl 100 × BSA and 25 μl *Sau*3AI. Incubate overnight at 37 °C in hot-air incubator.
2. Check for *Sau*3AI digestion: analyse 1 μl of uncut DNA, 1 μl and 3 μl of *Sau*3AI digestion (use bromophenol blue loading buffer and xylene cyanole loading buffer for uncut and *Sau*3AI-digested DNA, respectively) on a 3% agarose gel. Most (>80%) of the 110-bp fragment has been digested, and a faint band, corresponding to the ditags can now be detected at ~25 bp.
3. Add 125 μl xylene cyanole loading buffer to the digested DNA sample and load on a preparative 12% polyacrylamide vertical gel in 1 × TAE. Run at 30 mA until bromophenol blue of the size marker is 12 cm away from the well.
4. Transfer the gel in SYBR Green I stain at 1:10 000 dilution in 1 × TAE. Wrap the container in aluminium foil and stain gel for 20 min. Visualize on ultraviolet (UV) box.
5. Cut out the ditags band (24–26 bp) and transfer acrylamide slices in 0.5 ml tubes (for a 20-cm wide gel use eight tubes). Pierce the bottom of 0.5 ml tubes with a 18-gauge

Protocol 7 continued

needle. Place the tubes in 2 ml tubes and spin 5 min at 10 000 g. Prepare the following elution buffer for each tube: 475 μl LoTE, 25 μl 10 M ammonium acetate, 5 μg glycogen. Add 250 μl elution buffer in each 0.5 ml tube and centrifuge again. Discard 0.5 ml tubes and add 250 μl elution buffer directly in each 2 ml tube. Incubate overnight at 37 °C in hot-air incubator.

6. Prepare a series of 16 SpinX microcentrifuge tubes: add 20 μg glycogen in each collection tube. Transfer content of each 2 ml tube (~600 μl) to two SpinX microcentrifuge tubes. Spin 5 min at 13 000 g. Transfer 350 μl of eluted solution into 1.5 ml tubes (10–11 tubes), extract with PCI, perform high concentration ethanol precipitation. Wash twice with 75% ethanol, vacuum dry, and pool all pellets in 15 μl LoTE.
7. Measure the amount of purified ditags by dot quantitation (12) using 1 μl of sample. Total DNA at this stage is usually 1 μg, but a library can still be generated with 400 ng.
8. Ligate ditags to form concatemers: add to your sample (14 μl) 4.4 μl 5 × mix ligase, 2.2 μl 10 mM ATP, and 2.2 μl concentrated (5 U μl^{-1}) T4 DNA ligase.
9. Incubate 2 h at 16 °C. Stop the reaction by adding 5 μl of bromophenol blue loading buffer and store at −20 °C.

7.3 Purification of concatemers

We heat concatemers at 45 °C for 5 min immediately before loading on gel to separate unligated cohesive ends. Concatemers form a smear on the gel from about 100 bp to several kilobase pairs (*Figure 4*) and can be easily detected using SYBR Green I stain. All fragments >300 bp (i.e. with 25 or more tags) are potentially interesting for library construction. We usually cut out fragments of 350–600, 600–2000, and >2000 bp and generate a first library using 600–2000 bp DNA fragments. Longer fragments will be more informative but are expected to be cloned with poor efficiency.

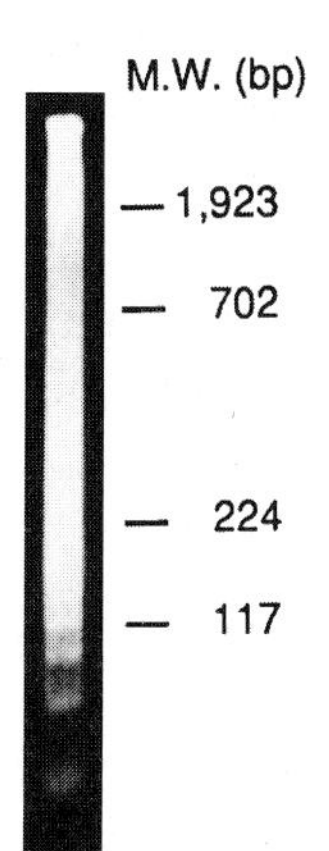

Figure 4 Gel analysis of concatemers. Ditags were concatenated by ligation (2 h at 16 °C), then electrophoresed through an 8% polyacrylamide gel. The gel was post-stained using SYBR Green I, and visualized by UV illumination at 305 nm. Migration of the molecular weight marker (λ *Bst*EII-digest) is indicated on the right.

Protocol 8

Purification and cloning of concatemers

Equipment and reagents

- SYBR Green I stain
- Vertical gel electrophoresis unit, with 20 × 20 cm plates, 1.5 mm thick spacers, and 20-well comb
- 8% polyacrylamide gel: 61.6 ml H_2O, 16 ml acrylamide 40% (37.5:1 acrylamide/bis), 1.6 ml 50 × TAE, 800 μl 10% ammonium persulphate, 69 μl TEMED
- 10 mM ATP
- 100-bp DNA ladder (Life Technologies, ref. 15628-019)
- T4 DNA ligase 1 U μl^{-1} (Life Technologies, ref. 15224-017) and 5 × reaction buffer
- pBluescript II, linearized with *Bam*HI and dephosphorylated
- *E. coli* XL2 Blue ultracompetent cells (Stratagene, ref. 200150)

Method

1. Heat sample 5 min at 45 °C and load into one lane of a 20 wells 8% acrylamide gel. Run at 30 mA until bromophenol blue is 10–12 cm from the well.
2. Stain the gel with SYBR Green I as described in *Protocol 7* and visualize on UV box.
3. Concatemers form a smear on gel with a range from about 100 bp to the gel well (*Figure 4*). Cut out regions containing DNA of 350–600, 600–2000, and >2000 bp. Purify separately DNA of each three slices as described in steps 5–6 of *Protocol 7* (a 1-h incubation period of gel slices in LoTE/ammonium acetate solution is sufficient). Resuspend the pellet in 6 μl LoTE and generate a first library using concatemers of 600–2000 bp.
4. Mix 6 μl of concatemers and 2 μl (25 ng) of *Bam*HI-cut pBluescript II. Heat 5 min at 45 °C then chill on ice.
5. Add 3 μl 5 × mix ligase, 1 μl H_2O, 1.5 μl 10 mM ATP, and 1.5 μl T4 DNA ligase (1 U μl^{-1}). Mix and incubate overnight at 16 °C.
6. Add 20 μg glycogen and 285 μl LoTE and extract with PCI. Ethanol precipitate, wash twice with 75% ethanol, vacuum dry, and resuspend the pellet in 12 μl LoTE.
7. Transform *E. coli* XL2 Blue ultracompetent cells with 1/3 (4 μl) of ligation reaction according to the manufacturer's instructions. Plate different volumes (5 μl, 10 μl, 20 μl, 40 μl) of transformation mix on to Petri dishes containing Luria agar supplemented with ampicillin, X-gal, and IPTG. Incubate 15–16 h at 37 °C. Save the remaining (~900 μl) transformation solution (add 225 μl 80% glycerol, mix intermittently for 5 min, and store at −80 °C). It will be used to plate additional bacteria if library appears correct.
8. Count insert-free (i.e. blue) and recombinant (i.e. white) bacterial colonies on each plate. The fraction of recombinant colonies should be >50%, and their total number should be in the range of 10 000–60 000 for 1 ml of transformation mix.

7.4 Cloning of concatemers

Concatemers can be cloned and sequenced in a vector of choice. We currently clone concatemers in pBluescript II linearized with *Bam*HI and dephosphorylated by calf intestinal alkaline phosphatase treatment. Any kind of vector with a *Bam*HI site in the multiple cloning site will be suitable. Velculescu *et al.* (7, 8) use pZero-1 from Invitrogen which only allows recombinants to grow (DNA insertion into the multiple cloning site disrupts a lethal gene). The competent cells and transformation procedures (heat shock or electroporation) can also be changed according to your facilities. Whatever your choice, it is important to use bacterial cells allowing very high cloning efficiency ($\geqslant 5 \times 10^9$ transformants μg^{-1} of supercoiled DNA). An important point is to evaluate the number of clones (*Protocol 8*, step 8) obtained in the library. As a large number (1000–2000) of clones will be sequenced, the total number of recombinants should be >10 000.

Library screening can be performed by PCR or DNA Miniprep. In our hands, DNA Miniprep provides more reproducible amounts of DNA than PCR, and avoids false positive signals. We use Qiaprep 8 Miniprep kit (Qiagen, ref. 27144) which enables to perform 96 minipreps in ~2 h. Plasmid DNA is eluted from Qiagen columns with 100 µl of elution buffer; 5 µl are digested to evaluate insert size, and if insert is >200 bp, 5 µl are directly used for DNA sequencing.

8 Data analysis

8.1 Software

Once cloned concatemers have been sequenced, tags must be extracted, quantified, and identified through possible data bank matches. Two software packages have been written for these purposes.

The first one was set up by K.W. Kinzler, and is generously provided upon request to academic investigators. SAGE software (7) was written in Visual Basic and operates on personal computers through the Microsoft Windows system. It extracts tags from text sequence files, quantifies them, allows to compare several libraries, and provides links to GenBank data downloaded from CD-ROM flat files or over the Internet. The latter function enables rapid identification of tags originating from characterized genes or cDNAs. However, description of EST sequences is truncated, which constrains one to looking for individual GenBank reports. SAGE software also includes several simulating tools which allow, for example, assessment of the significance of differences observed between two libraries, and evaluation of sequencing accuracy.

The second software is currently developed at the University of Montpellier-2 (France) by J. Marti and co-workers, as part of a database (CbC, for Cell by Cell) intended to store and retrieve data from SAGE experiments. Scripts for extraction of data are developed in C language under Unix environment and the database management system implemented in Access®. Text files are concatenated to yield the working file from which tag sequences are extracted and enumerated. Treatment of raw data involves identification of vector contaminants, truncated and repeated ditags (see below). For experiments on human, mouse, and rat cell samples, the tags are searched in the non-redundant set of sequences provided by the UniGene collection. These data can be loaded from the anonymous FTP site: ncbi.nlm.gov/repository/unigene/. Useful files are Hs.data.Z and Hs.seq.uniq.Z for humans, and similar files for mouse and rat. The results are displayed in a table which provides the sequence of each tag, its number of occurrences with the matching cluster number (Hs for *homo sapiens*, Mm for *Mus musculus*, and Rn for *Rattus norvegicus*), and other data extracted from the source files, including GenBank accession numbers. For human genes, when available, a link is automatically established with GeneCards, (http://bioinfo.weizmann.ac.il/cards/) allowing one to get additional information.

8.2 Library validity

8.2.1 Insert length

Initial assessment of library quality will be obtained from screening for insert length. A good library should contain >60% clones with inserts >240 bp (20 tags). Only one DNA strand is sequenced, as accuracy is obtained from the number of tags recorded, rather than from the quality of individual runs. Depending on your budget and sequencing facilities, either all clones or only the most informative ones will be sequenced. It should be noted that the average length of inserts does not fit with that of gel-purified concatemers. Although we usually extract 600–2000 bp long concatemers, most of the clones have inserts <600 bp, and we never get inserts >800 bp. A number of reasons can explain such a paradoxical result. Indeed, long inserts are known to be cloned with poor efficiency. In addition, they are expected to contain several repeats or inverted repeats, and may thus form unstable plasmid constructs. Supporting this interpretation, it has been demonstrated (14), and we have also observed, that efficient removal of linkers (which represents up to 20% of total tags in poor libraries) increases the average length of cloned inserts. At any rate, it is worth emphasizing that similar biological information is obtained from libraries with short and long inserts.

8.2.2 Gene expression pattern

The basic pattern of gene expression in eukaryotic cells has been established long ago by kinetic analysis of mRNA-cDNA hybridization (15, 16). In a 'typical ' mammalian cell, the total RNA mass consists of 300 000 molecules, corresponding to ~12 000 transcripts which divide into three abundance classes. A very small number of mRNAs (~10) are expressed to exceedingly high levels (3000–15 000 copies cell^{-1}). A larger number of mRNAs (~500) reach an expression level in the range of 100–500 copies cell^{-1}. Finally, the majority of mRNAs (>10 000) are poorly expressed (10–100 copies cell^{-1}). This basic pattern should be observed in SAGE or SADE libraries. However, before translating tags abundance in a definite gene expression profile, the data must be scrutinized for artefacts encountered in library construction.

8.2.3 Occurrence of linker-derived sequences

As mentioned above, some libraries display a high amount of linker sequences. If this amount is 20% or more, sequencing will be quite expensive, and it is better to start again from the RNA sample. Library contamination with 10–15% of linker sequences is acceptable, 5–10% is good, and <5% is excellent. In addition to the two perfect linker matches (GTCCCTGTGC and GTCCCTTCCG), reading ambiguities can lead to sequences with one mismatch. These linker-like sequences are also easily identified as, assuming efficient enzymatic cleavage, the probability of having adjacent *Sau*3AI and *Bsm*FI sites in the concatemers is normally zero. Linker and linker-like sequences can be automatically discarded using SAGE or CbC software, and their relative amounts can be used to evaluate the sequencing accuracy (see Section 8.1).

8.2.4 Duplicate ditags

Another category of sequences that must be deleted are those corresponding to duplicate ditags. Indeed, except for peculiar tissues (e.g. lactating gland or laying hen oviduct) in which one or a very small number of transcripts constitutes the bulk of the mRNA mass, the probability for any two tags to be found several times in the same ditag is very small. Elimination of repeated ditags will therefore correct for preferential PCR amplification of some targets, and for picking several bacterial colonies originating from the same clone. Most ditags (>95%) generally occur only once when the library is constructed from macroamounts of tissue. For microlibraries, the percent of unique ditags is generally lower. When it is no longer compatible (<75%) with efficient data acquisition, it is recommended to start again from the first (small-scale) PCR (see *Protocol* 6). Duplicate ditags are automatically retrieved from the sequence files by SAGE and CbC softwares.

8.3 Number of tags to be sequenced

The number of tags to be analysed will obviously depend on the application and tissue source. As a matter of fact, reducing the tissue complexity through isolation of defined cell populations will allow to diminish markedly the minimum number of tags for accurate analysis, and to better correlate molecular and physiological phenotypes.

Delineation of the most expressed genes (>500 copies cell^{-1}) in one tissue, and comparison with their expression level in another one, will require to sequence only a few thousand tags (*Figure* 5). Analysing 5000 tags, 300 will be detected at least three times. As automated sequencers can read 48–96 templates simultaneously, 10 000 tags will be recorded from five to ten gels if the average number of tags per clone is ~20.

The most difficult projects are those aiming to compare gene expression profiles in the same tissue under two physiological or pathological conditions. Differentially expressed genes could belong to any of the three abundance classes and, furthermore, they can be either upregulated or downregulated. A reasonable

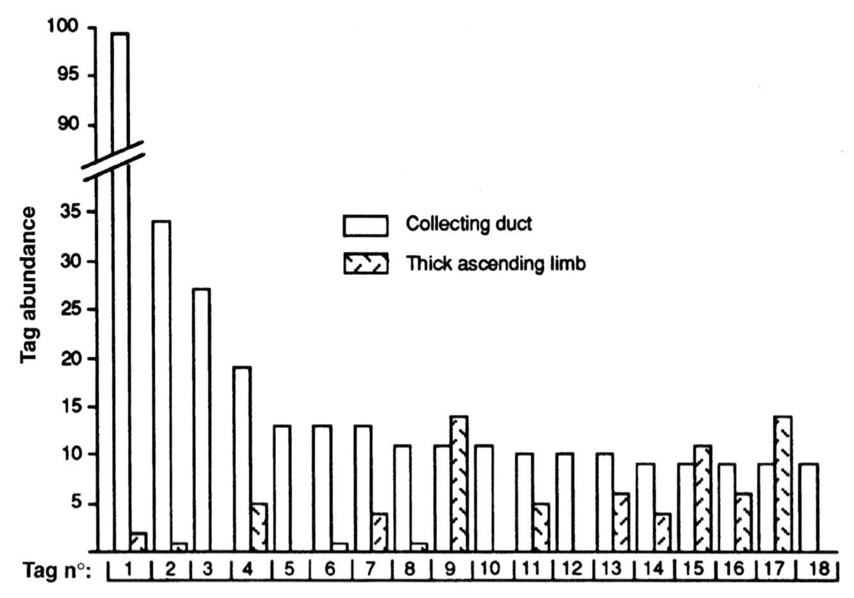

Figure 5 Comparison of gene expression levels in two nephron portions of the mouse kidney. SADE libraries were constructed from ~50 000 cells isolated by microdissection from medullary collecting ducts or medullary thick ascending limbs, and 5000 tags were sequenced in each case. The data show the 18 most abundant collecting duct tags originating from nuclear transcripts (mitochondrial tags were excluded from the analysis), and their corresponding abundance in the thick ascending limb library.

number of tags to be sequenced would be in the range of 30 000–50 000. The probability (p) of detecting a sequence of a given abundance can be calculated from the Clarke and Carbon (17) equation ($N = \ln(1 - p)/\ln(1 - x/n)$), where N is the number of sequence analysed, x is the expression level, and n is the total number of mRNAs per cell (~300 000). Thus, the analysis of 30 000 and 50 000 tags will provide a 95% confidence level of detecting transcripts expressed at 30 and 18 copies cell^{-1}, respectively. Most upregulation processes will be therefore assessed. For example, tags corresponding to poorly expressed transcripts may be detected 1 and ⩾5 times in control and experimental conditions, respectively. However, we have to be aware that the possibility of assessing downregulation processes will be less exhaustive. It will only concern tags present more than five to ten times in the control condition, which excludes from the analysis part of the poorly expressed transcripts.

References

1. DeRisi, J. L., Iyer, V. R., and Brown, P. O. (1997). *Science*, **278**, 680.
2. Wodicka, L., Dong, H., Mittmann, M., Ho, M. H., and Lockhart, D. J. (1997). *Nature Biotechnol.*, **15**, 1359.
3. Gress, T. M., Hoheisel, J. D., Lennon, G. G., Zehetner, G., and Lehrach, H (1992). *Mamm. Genome*, **3**, 609.
4. Piétu, G., Alibert, O., Guichard, V., Lamy, B., Bois, F., *et al.* (1996). *Genome Res.*, **6**, 492.

5. Adams, M. D., Kerlavage, A. R., Fleischmann, R. D., Fuldner, R. A., Bult, C. J., *et al.* (1995). *Nature*, **377**, 3.
6. Okubo, K., Hori, N., Matoba, R., Niiyama, T., Fukushima, A., *et al.* (1992). *Nature Genet.*, **2**, 173.
7. Velculescu, V. E., Zhang, L., Vogelstein, B., and Kinzler, K. W. (1995). *Science*, **270**, 484.
8. Velculescu, V. E., Zhang, L., Zhou, W., Vogelstein, J., Basrai, M., *et al.* (1997). *Cell*, **88**, 243.
9. Zhang, L., Zhou, W., Velculescu, V. E., Kern, S. E., Hruban, R. H., *et al.* (1997). *Science*, **276**, 1268.
10. Polyak, K., Xia, Y., Zweier, J. L., Kinzler, K. W., and Vogelstein, B. (1997). *Nature*, **389**, 300.
11. Sade, D. A.F. (1990). Oeuvres complètes. Gallimard, Paris.
11b. Virlon *et al.* PNAS 96: 15286 (1999).
12. Ausubel, F. M., Brent, R., Kingston, R. E., Moore, D. D., Seidman, J. G., *et al.* (1993, updated quaterly). *Current protocols in Molecular Biology*. Greene Publishing Associates and Wiley-Interscience, New York.
13. Chomczynski, P. and Sacchi, N. (1987). *Anal. Biochem.*, **162**, 156.
14. Powell, J. (1998). *Nucleic Acids Res.*, **26**, 3445.
15. Hastie, N. D. and Bishop, J. O. (1976) *Cell*, **9**, 761.
16. Hereford, L. M. and Rosbash, M. (1977). *Cell*, **10**, 453.
17. Clarke, L. and Carbon, J. (1976). *Cell*, **9**, 91.

Chapter 8

Analysis of gene expression in single cells using three prime end amplification PCR

Tom C. Freeman
Sanger Centre, Cambridge, UK

Alistair K. Dixon
Parke Davis Neuroscience, Cambridge, UK

Kevin Lee
Parke Davis Neuroscience, Cambridge, UK

Peter J. Richardson
University of Cambridge, UK

1 Introduction

Many tissues and biological systems are composed of highly heterogeneous cell populations. This makes it difficult to correlate the functional characteristics or composition of a system as a whole with the activity and molecular profile of individual cell types. In the central nervous system for example, many are attempting to dissect the biochemical basis of cell function and unravel the complex interactions that take place between neurons. In cancer, there is a great need to examine the specific changes that occur in the malignant cells in isolation of the surrounding 'normal' non-transformed cell population, in order to understand the progression of the disease and identify potential therapeutic targets. The ability to analyse gene expression in single or small populations of cells provides a powerful tool to identify the differences between one cell type and another, and correlate function with the presence of specific gene products.

There are a number of approaches available for studying the expression of genes at the cellular level. *In situ* hybridization can be very informative when you wish to analyse the expression pattern of a gene on a tissue section, but it is only capable of analysing one or two genes in any given cell at a time. In addition, it is often too insensitive to detect low abundance mRNA species. A number of techniques have therefore been developed that use the polymerase chain reaction (PCR) to analyse gene expression in individual cells. 'Nested' PCR has been used

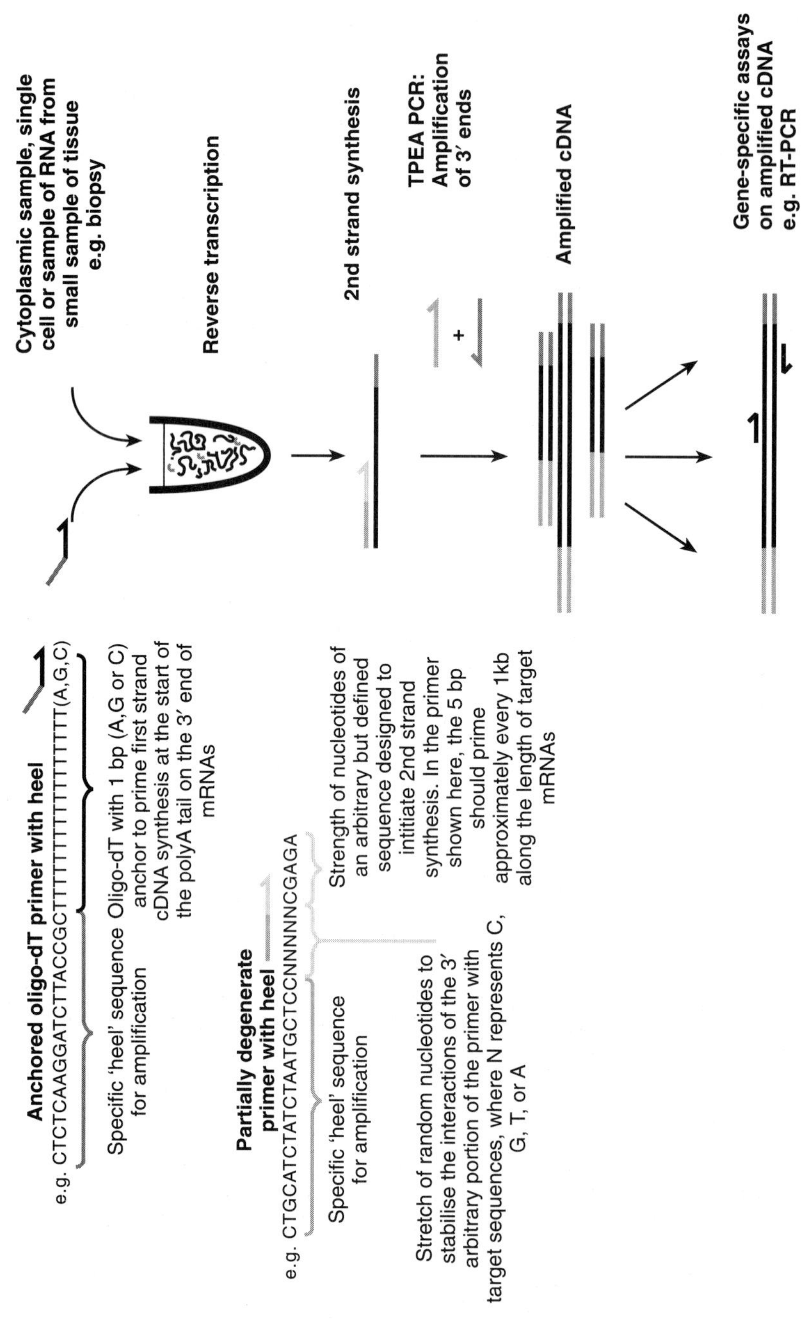

Figure 1 Schematic of TPEA PCR.

as a means of obtaining sufficient amplification and specificity for particular mRNA species to be detected in single cells (1–3). While this approach has allowed functional properties of identified neurons to be equated with the genes expressed (for examples, see refs 4 and 5), nested PCR restricts the analysis to small groups of closely related genes. Alternatively, the technique of RNA amplification permits numerous different mRNAs to be detected in samples of single cells (for example, ref. 6). This approach, however, is demanding technically, involving many manipulations (with the associated risks of loss of material and contamination), which also limits sample throughput.

In this chapter we describe our experiences of performing single cell analysis. We have employed two approaches to collect single cell material, fluorescence-activated cell sorting of cultured cells and the collection of the cytoplasm from single neurons and coupled this with a novel technique, called three prime end amplification (TPEA) PCR (7), which has allowed us to examine the expression of a large number of diverse genes in single cells.

2 Cytoplasmic harvesting

2.1 Collection of cytoplasm from fluorescence-activated cell sorter (FACS) sorted cells

This procedure is applicable to any type of dissociated or free-living cells that can be specifically labelled with a fluorescent tags (e.g. fluorescently labelled antibodies to surface antigens), allowing specific subpopulation of cells to be isolated. In order to develop the method of TPEA PCR, an Epstein–Barr virus transformed lymphoblastoid cell line maintained in log phase was used a convenient source of single cell material (7). Cells were labelled with a fluorescent dye which incorporates into nuclear DNA to facilitate the FACS sorting procedure and sorted directly into a lysis buffer containing the detergent NP-40. The lysis buffer disrupts the plasma membrane but leaves nuclear membranes intact (8) This allows the nucleus to be removed by centrifugation prior to reverse transcription and therefore prevents contamination of the sample with DNA and nuclear RNA.

Protocol 1

Lysis of FACS sorted cells

Equipment and reagents

- 5 × Lysis buffer: 25 μl 1 M Tris-HCl; •37.5 μl 1 M KCl; 1.5 μl 1 M $MgCl_2$; 2.5 μl NP-40; 33.5 μl DEPC-treated water
- RNA guard solution (to be prepared fresh on the day of the experiment): 6 μl 5 × first strand buffer (Gibco BRL); 6 μl RNase inhibitor (35 U μl^{-1}; Pharmacia); 18 μl DEPC-treated water
- Lysis solution—sufficient for 45 cells (to be prepared fresh on the day of the experiment): 75 μl of 5 × lysis buffer; 25 μl of RNA guard mix; 250 μl DEPC-treated water

Protocol 1 continued

Cell sorting and lysis

1. Pipette 7.5 μl of lysis solution into wells of a 96-well plate (Costar).
2. Stain cells with appropriate dye to enable sorting of the target cell population and FACS sort individual cells into wells of the plate containing lysis solution. Monitor sorting procedure to ensure that the majority of wells contain the desired number of cells.[a]
3. Incubate the cells on ice for 5 min.
4. Transfer the contents of individual wells into 0.5 ml microfuge tubes and centrifuge at 8000 g for 5 min at 4 °C.[b]
5. Carefully aspirate the supernatant. In order not to disturb the pellet take only 5 μl leaving the remaining 2 μl at the site where the nucleus should have pelleted.
6. Transfer the cell lysate containing the RNA to a sterile 0.5 ml tube and proceed to reverse transcription.

[a] The precise conditions depend on the FACS machine being used, the cells to be sorted and the labelling method. We labelled all the cells in a population with the dye Hoecsht 33342 (Sigma; 1 μg ml^{-1} for 30 min at 37 °C) and sorted using the Autoclone attachment of a Coulter Elite ESP flow cytometer. Time of flight, forward and right angle scatter, and fluorescence peak and area measurements were used to ensure the efficient sorting of single cells.

[b] Note that the cells were not sorted directly into these tubes, due to their angled walls on which the cells could land and so miss the lysis solution.

2.2 Isolation of neuronal cytoplasm from single neurons

2.2.1 Equipment and preparation

Extensive precautions need to be taken to minimize the risk of sample contamination by RNases and nucleic acids. Our experience of examining gene expression in striatal cholinergic neurons is that by taking the precautionary measures outlined in *Protocol 2*, a good level of success can be achieved using this methodology. For successful collection of the cytoplasmic contents of neurons within brain slices, it is essential to have high-quality visual control of the process and good mechanical control of the harvesting electrode. We use the 'Infrapatch' system (Luigs and Neumann), which uses infrared light to view neurons within brain slices (9). This system comprises a Zeiss Axioskop microscope fitted with a ×64 objective lens together with independent XYZ manipulators for microscope, electrode and bath chamber. Our work routinely uses 300 μm brain slices from 14 to 28-day old rats prepared using a vibratome at 4 °C. The slices are stored in oxygenated cerebrospinal fluid (CSF) (see *Protocol 2*) at room temperature prior to use.

Protocol 2

Preparation for harvesting neuronal cytoplasm

Equipment and reagents

- 1 M NaOH
- DEPC-treated water
- 10 × stock of artificial CSF: 1.25 M NaCl; 25 mM KCl; 12.5 mM NaH_2PO_4; 20 mM $CaCl_2$; 10 mM $MgCl_2$
- $NaHCO_3$ (25 mM final concentration)
- Glucose (10 mM)
- 95%, 5% O_2/CO_2
- Vacuum filter
- Pre-autoclaved sintered glass bubbler
- 10 × stock electrode buffer: 1.2 M K-gluconate; 100 mM NaCl; 20 mM $MgCl_2$; 5.0 mM K_2EGTA; 100 mM HEPES; pH to 7.2 with KOH
- Na_2ATP (Pharmacia)
- Na_2GTP (Pharmacia)
- Dithiothreitol (DTT) (1 mM; Gibco BRL)
- Glycogen (0.2 mg ml^{-1}; Boehringer)
- RNasin (0.4 U ml^{-1}; Promega)

All reagents must be prepared in DEPC-treated water and all glassware autoclaved.

Equipment preparation

1. Clean the perfusion tubing with 1 M NaOH and flush thoroughly with DEPC-treated water.
2. Use a sterile 1 ml syringe to provide pressure or suction to the electrode via pre-autoclaved tubing.
3. Bake electrode glass (200 °C for 4 h) individually wrapped in aluminium foil.
4. Thoroughly clean the electrode holder with ethanol on the day of experiment.

Solution preparation

10 × stock of artificial CSF (aCSF):

- 1.25 M NaCl; 25 mM KCl; 12.5 mM NaH_2PO_4; 20 mM $CaCl_2$; 10 mM $MgCl_2$
- Dilute appropriately on the day of the experiment with DEPC-treated water adding $NaHCO_3$ (25 mM final concentration) and glucose (10 mM)
- Filter through a 0.22 μm vacuum filter and oxygenate with 95%, 5% O_2/CO_2 using a pre-autoclaved sintered glass bubbler

10 × stock electrode buffer:

- 1.2 M K-gluconate; 100 mM NaCl; 20 mM $MgCl_2$; 5.0 mM K_2EGTA; 100 mM HEPES; pH to 7.2 with KOH
- Prior to DEPC treatment, sterile filter (0.22 μm filter) stock solution to remove particulate matter
- Dilute to final concentration on the day of the experiment, adding Na_2ATP,[a] Na_2GTP,[a] DTT (1 mM), glycogen[b] (0.2 mg ml^{-1}), and RNasin[c] (0.4 U ml^{-1})

[a] Routinely we add adenosine triphosphate (ATP) to a final concentration of 4 mM, and guanosine triphosphate (GTP) to 0.3 mM.

[b] The glycogen should be filtered before use (0.22 μm filter) to prevent blockage of the electrode. The glycogen is included in the electrode buffer to reduce the retention of RNA by the electrode walls.

[c] RNasin is supplied in 40% glycerol and therefore can affect the osmolarity of the electrode solution if added in excess of the quantities recommended.

Although extensive precautions need be taken when undertaking this single cell procedure, once the equipment is clean it takes little effort to maintain.

2.2.2 Seal formation and cytoplasm harvesting

A great advantage of using brain slices is that the functional properties of specific neurons can be assayed and these measurements can be used to confirm the identity of the neuronal cell type, prior to harvesting a cell's cytoplasm. However, there is an increased risk of contamination of the electrode by broken cells within the slice. For this reason visual control is a great advantage and the maintenance of positive pressure in the electrode during the approach to the cell is essential.

When harvesting cell contents, it is important to apply suction gradually and to visualize the slow removal of cytoplasm from the cell. The rapid application of strong suction usually results in the nucleus being pulled across the cell, plugging the electrode and therefore preventing further harvesting. It has been found that moving the electrode around the cell slightly (i.e. away from the nucleus) helps prevent blockage. At all times, however, it is important to monitor electrically the integrity of the seal.

Protocol 3

Seal formation and harvesting

Equipment and reagents

- 70% ethanol
- DEPC-treated water
- 2–4 μl of electrode solution
- Reverse transcription buffer

Clean gloves should be worn at all times and all surfaces cleaned with 70% ethanol, and the environment of the brain slice maintained as clean as possible. Repeat the following steps for each cell harvesting.

Method

1. Draw the glass electrode to give a resistance of approximately 4 MΩ, and polish the tip.
2. Rechloride the electrode wire with bleach, rinse away excess bleach using DEPC-treated water.
3. Fill the electrode with 2–4 μl of electrode solution using a fresh 25G needle[a] for each cell.
4. Mount the electrode and, under visual control, manoeuvre into the bath solution towards the cell with positive pressure applied at all times.
5. Approach the chosen cell, maintaining the positive pressure, and form a 'cell attached' seal. Apply suction to obtain the 'whole cell' configuration.[b]

Protocol 3 continued

6 Apply gentle suction and watch the cell contents passing into the electrode tip. On average this takes about 10 min.

7 Apply strong suction to pull the nucleus onto the electrode tip, giving an electrical seal of approximately 1 GΩ. Withdraw the electrode rapidly from the slice, and use a 50 ml syringe to force the electrode contents into reverse transcription buffer.[c]

[a] We use a Gilson pipette fitted with a sterile 25G needle to fill electrodes with electrode solution. The cut-off end of a sterile plugged pipette tip is used to provide an adaptor between the pipette and needle. Each time the electrode is filled, the needle and cut off tip should be changed.

[b] When forming a seal, it is important to try and place the electrode away from the nucleus which can be visualized as one focuses through the cell.

[c] This process is performed using a sterile 50 ml syringe to which is attached an autoclaved piece of silicone tubing of a diameter that allows the electrode to be tightly inserted. Positive pressure is applied to expel the electrode contents and the tip of the electrode broken into the microtube under reverse transcription buffer (see *Protocol* 4).

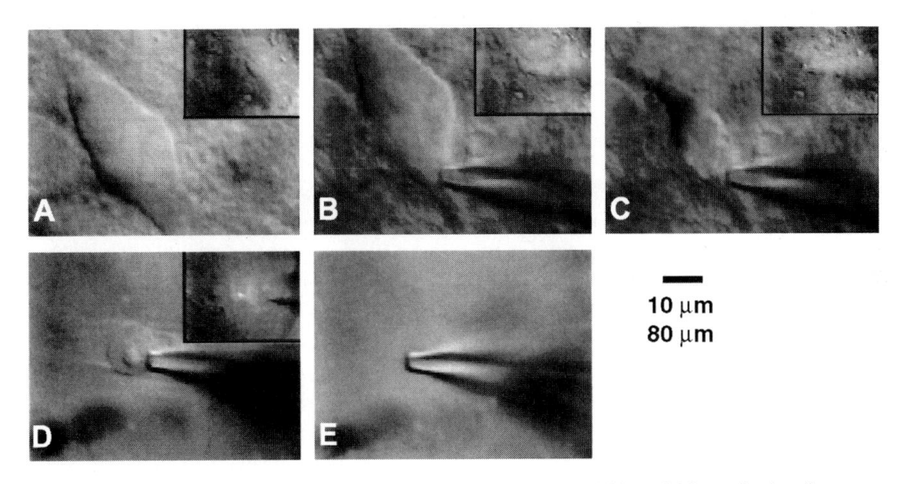

Figure 2 Harvesting of cell contents from a neuron situated within a 300 μm brain slice. (a) The location of the neuron in the brain slice. (b) The patch clamp electrode has been placed on the surface of the neuron and the whole cell recording configuration formed. (c) The cytoplasmic contents have been harvested while in (d) the electrode has been withdrawn away from the neuron and out the brain slice with subsequent formation of a nucleated patch. (e) Withdrawal of the nucleus through the air–electrolyte interface results in loss of integrity of the nucleated patch.

It is normally possible to remove up to approximately 40% of the somatic cell cytoplasm. Once the harvesting process is complete, others have tended to use the outside-out patch formation as a means of retaining the harvested contents within the electrode while preventing contamination following pipette withdrawal. In our experience the formation of outside-out patches in harvested cells where a large amount of the cytoplasmic skeleton has been damaged is difficult and is associated with a low success rate (approximately 20% of attempts). To overcome this problem, we tend to use nucleated patches, i.e. once the cytoplasm has been harvested we apply strong suction which results in the nucleus attaching to the electrode tip. The continued application of strong suction in concert with pipette withdrawal results in the formation of a nucleated patch with seal resistance often in excess of 1 GΩ, although seal resistances of 500 MΩ are also usable. Using this approach we have achieved a success rate (i.e. detection of expression of six housekeeping genes in the absence of contamination) of approximately 80% harvested cells. The level of contamination associated with this procedure is no more than that seen with outside-out patch formation and we have never detected contamination with nuclear contents with either procedure. Appropriate controls for this stage include taking the electrode up to a cell without forming a seal, as well as samples of the aCSF passing the electrode.

3 TPEA PCR

3.1 Reverse transcription

The reverse transcription is performed using an anchored oligo(dT) primer designed to prime reverse transcription at the start of the poly(A) tail, which also bears a unique heel sequence at the 5′ end that will serve later as a priming site for TPEA PCR.

Protocol 4

Reverse transcription

Equipment and reagents

2.5 × reverse transcription buffer

- 50 μl 5 × First strand buffer (Gibco BRL, supplied with MMLV reverse transcriptase)
- 25 μl 100 mM DTT (Gibco BRL, supplied with MMLV)
- 6.25 μl 10 mM dNTPs (Boehringer)
- 10 μl RNasin (Promega)
- 0.125 ng μl^{-1} reverse transcription primer (final concentration). Reverse transcription primer: CTCTCAAGGATCTTACCGC$(T)_{19}$(A,G,C)(T,A,G,C)
- DEPC-treated water

200 Units MMLV reverse transcriptase (in 2 μl)

Method

1 Prepare a 2.5 × reverse transcription buffer on the day of the experiment:
   - 50 μl 5 × First strand buffer (supplied with MMLV reverse transcriptase)
   - 25 μl 100 mM DTT (supplied with MMLV)

Protocol 4 continued

- 6.25 μl 10 mM dNTPs
- 0.125 ng μl^{-1} reverse transcription primer (final concentration). Reverse transcription primer: CTCTCAAGGATCTTACCGC$(T)_{19}$(A,G,C) (T,A,G,C)
- 10 μl RNasin
- Add DEPC-treated water to a total volume of 100 μl

2a (Sorted cells) Add 5 μl of cell cytoplasm in lysis buffer to 5 μl of 2.5 × reverse transcription buffer.

2b (Neuronal cytoplasm) Expel contents of electrode[a] under pressure by breaking the electrode tip under 4 μl of 2.5 × reverse transcription buffer.[b]

3 Heat to 65 °C for 2 min (to remove secondary structure of RNA), place on ice, add 200 Units of MMLV reverse transcriptase (in 2 μl) and incubate at 37 °C for between 60 and 90 min.

4 Store frozen at −20 °C until used.

[a] When harvesting using patch clamp electrodes, the volume within the electrode is not easy to control. We have however found that the reverse transcription protocol outlined below is not sensitive to the variation in volume usually produced by this procedure, i.e. between 2 and 4 μl.

[b] See *Protocol 3*.

3.2 Second strand synthesis and TPEA PCR

TPEA PCR uses a novel forward primer that has been designed to initiate second strand synthesis along the length of the cDNA generated during reverse transcription. In order to perform this task the primer contains (5′–3′): a unique 20-mer heel sequence (CTGCATCTATCTAATGCTCC), followed by a stretch of five random nucleotides (NNNNN). In addition the primer contains a defined pentameric sequence at the 3′ end. The sequence of these five bases can be varied according to requirements, we have routinely used CGAGA, or a mixture of four different sequences (CGAGA, CGACA, CGTAC, and ATGCG). The way that this primer is envisaged to work is that the 3′ pentameric sequence binds arbitrarily to the first strand cDNA approximately every 1 kb (4^5), this interaction being stabilized by the stretch of five random bases that precede it, allowing Taq polymerase to begin second strand synthesis. In subsequent rounds of PCR the 5′ heel provides a specific priming site that can be used in conjunction with the heel sequence incorporated during reverse transcription allowing the simultaneous amplification of the 3′ portion of all cDNA species present in the sample. Indeed, the amplification of only the 3′ portion of a gene's sequence is seen as a desirable feature, as it circumvents the bias against longer sequences that occurs when trying to amplify populations of DNA comprised of molecules of very different lengths.

Protocol 5

Second strand synthesis

Second strand primer CTGCATCTATCTAATGCTCCNNNNNCGAGA

Equipment and reagents

- 10 × PCR buffer: 670 mM Tris-HCl pH 8.3, 45 mM $MgCl_2$, 166 mM ammonium sulphate
- 1/10 bovine serum albumin (Sigma 5%)
- 1/20 β-mercaptoethanol
- 10 mM dNTPs
- Amplitaq (5 U μl^{-1}; Perkin-Elmer)
- Second strand primer(s) (10 pg/μl^{-1})
- Reaction mixes for the TPEA amplification
- Taq enzyme

Method

1 Prepare the following reaction mix, but do not add the Taq enzyme until immediately before use.

Mix 1	Per 10 reactions
10 × PCR buffer (670 mM Tris-HCl pH 8.3, 45 mM $MgCl_2$, 166 mM ammonium sulphate)	8.33 μl
1/10 bovine serum albumin (5%)	4.95 μl
1/20 β-mercaptoethanol	2.1 μl
10 mM dNTPs	7.5 μl
Amplitaq (5 U μl^{-1})	1.25 μl
Second strand primer(s) (10 pg/μl^{-1})	10 μl
H_2O	15.87 μl

2 Add 5 μl of Mix 1 to the reverse transcribed contents (10 μl) derived from one cell and subject to the following protocol: 92 °C for 2.5 min; 50 °C for 7.5 min; 72 °C for 8 min; 10 °C for 1 min.

3 While the second strand synthesis is taking place, the reaction mixes for the TPEA amplification should be prepared, care being taken not to mix the Taq enzyme with the 3′ and 5′ primers until the last moment.

Protocol 6

TPEA PCR

Equipment and reagents

- Mix 2: 10 × PCR buffer; 1/10 BSA; 1/20 β-mercaptoethanol; 10 mM dNTPs; Taq
- Mix 3: Mix 2; 1 ng μl^{-1} 3′ primer; 1 ng μl^{-1} 5′ primer
- 10 mM Tris/0.1 mM EDTA
- Mix 4: Mix 2; 10 ng μl^{-1} 3′ primer; 10 ng μl^{-1} 5′ primer
- 5′ primer: CTGCATCTATCTAATGCTCC
- 3′ primer: CTCTCAAGGATCTTACCGC

Protocol 6 continued

Method

1 Prepare the TPEA reaction mixes, add primers immediately before use:

	For 10 reactions
(a) Mix 2	
10 × PCR buffer	25 µl
1/10 BSA	8.25 µl
1/20 β-mercaptoethanol	3.5 µl
10 mM dNTPs	12.5 µl
Taq	1.5 µl
(b) Mix 3	
Mix 2	10.16 µl
1 ng μl^{-1} 3′ primer	10 µl
1 ng μl^{-1} 5′ primer	10 µl
H_2O	19.84 µl
(c) Mix 4	
Mix 2	20 µl
10 ng μl^{-1} 3′ primer	10 µl
10 ng μl^{-1} 5′ primer	10 µl
H_2O	160 µl
Total	200 µl

2 After the second strand synthesis add 5 µl of Mix 3 and subject to 10 rounds of PCR: 1 × 92 °C for 2 min; 10 × 92 °C for 0.5 min; 10 × 60 °C for 1.5 min; 10 × 72 °C for 1 min; 1 × 72 °C for 10 min.

3 After this initial amplification add 20 µl Mix 4 and subject to a further 40 rounds of amplification using the same cycling conditions as in step 3.

4 Dilute the final reaction mixture to 200 µl with 10 mM Tris/0.1 mM EDTA (pH 8.1).

4 Gene-specific PCR

We use gene-specific PCR after TPEA PCR to assess which genes were expressed in the original sample. It should be possible to assay for the expression of at least 40 genes in a sample derived from a single cell. The exact number of genes that can be analysed will depend on which genes are of interest; highly expressed genes, e.g. actin, ribosomal proteins, etc. require very much less than 2.5% of the final TPEA PCR product as a template for gene-specific PCR, while low abundance transcripts may require more. The efficiency of the gene-specific assays will also influence the amount of TPEA PCR product required to detect a particular gene.

4.1 Practical considerations

The power of PCR lies in its ability to amplify specifically from a very small amount of a DNA template. Just a few copies of a sequence is all that is normally necessary to generate microgram or milligram amounts of the amplicon (the DNA sequence spanned by and including the primers). The exquisite sensitivity of PCR can, however, potentially cause problems. For many applications in expression analysis it is the presence or absence of a particular amplicon at the end of the PCR reaction that will used to assess whether a gene is expressed in a sample. Therefore, if any of the solutions or reagents become contaminated with even small amounts of 'foreign' DNA problems will arise in interpreting the results correctly. For this reason measures must be taken to avoid this happening and controlling for it in case it does. The laboratory should be maintained to a high degree of cleanliness, surfaces and equipment being decontaminated at regular intervals. Of particular importance is the need to keep the PCR set-up separate from the analysis of products. So if possible the two operations should take place in different areas of the laboratory and use separate equipment for each. Use plugged pipette tips for setting up the PCR and store stock solutions in small aliquots to avoid repeated use of the same pot of reagent.

4.2 Primer design and testing

The majority of our primers used to analyse for the presence of specific transcripts following TPEA PCR are designed to the 3′ portion of each gene sequence, preferably within 300 bp of the start poly(A) tail (where this is known). The 3′ UTR is generally the most heterogeneous (specific) portion of a gene's sequence, closely related genes often showing a high degree of homology in the coding region, and the portion of the sequence most likely to be represented in the TPEA PCR product. It is advisable to design primers to this region of the target sequence, although more 5′ primers also often work equally well. For this reason we also tend to design small amplicons, between 120 and 250 bp in length, although primers that generate larger products may be suitable. In the absence of genomic DNA in the RNA used to generate the cDNA template, it is possible to design primers to any 3′ portion of the RNA. However, if the genomic structure of a gene of interest is known, it can sometimes be desirable to design primers that span an intron and therefore give a mRNA-specific amplicon. We use the primer design program PRIMER (Whitehead Institute, Cambridge, MA) to select 20 ± 2 bp primers from this region that have a GC content of between 20 and 80% and a melting temperature (T_m) of 60 °C ± 3 °C. Using various cDNAs as templates, we have found that approximately 85% of primers designed to these criteria and run under the conditions described in *Protocol 7*, give a single band of the expected size with no optimization other than using either 60 or 55 °C as the annealing temperature.

It is important to test all primers prior to use for single cell analysis. All primers must be shown to generate a single PCR product of the expected size. They should be able to detect a given transcript in cDNA from the tissue or culture from which the cellular material has been derived (if they give a signal at the cellular level) and they should be able to detect low copy numbers of transcripts, i.e. provide a sensitive assay. One way to test the sensitivity of an assay is to purify the amplicon, then prepare a dilution series of it down to very low copy numbers and reamplify using PCR. Good primers should be able to amplify a visible amount of product at least down to 100 copies of the amplicon, and many down to as low as 10 copies.

Protocol 7

Gene-specific PCR

Stock solutions

For consistent PCR results prepare in advance sufficient solutions to complete a given study.

1 Cresol red solution:

- 84.5 mg cresol red sodium salts (Aldrich) in 100 ml T0.1E (10 mM Tris, 0.1 mM EDTA, pH 7.6)
- Store as 10 × 10 ml aliquots at -20 °C

2 10 × PCR reaction buffer:

- 1 M Tris-HCl pH 8.8 45 ml
- Cresol red solution 50 ml
- H_2O 1.5 ml
- 1 M $MgCl_2$ (3.5 mM final) 3.5 ml
- $(NH_4)_2SO_4$ (Gibco BRL) 1.454 g
- Store as 50 × 2 ml aliquots at −20 °C

3 Dilution buffer:

- H_2O 100 ml
- T0.1E 50 ml
- Cresol red solution 0.8125 ml
- 4 M NaOH (to pH) 50 μl
- Store as 48 × 3.1 ml aliquots at −20 °C

4 34.6% Sucrose:

- 121.1 g in 350 ml H_2O
- Store as 7 ml aliquots at −20 °C

5 10 mM dNTP:

- Ultra pure dNTP set (Pharmacia)
- 4 × 250μl 100 mM dATP
- 4 × 250μl 100 mM dCTP
- 4 × 250μl 100 mM dGTP
- 4 × 250μl 100 mM dTTP
- 6 ml H_2O
- Store as 1 ml aliquots at -20 °C

6 DNA standards

- Mouse genomic DNA (Clontech)
- Human placenta genomic DNA (Sigma)
- Make up to 4 ng μl^{-1}

Protocol 7 continued

PCR

1. Remove stocks, primers and cDNAs from freezer and allow to thaw.
2. Prepare an appropriate amount of the following master mix:

	Per reaction (μl)
34.6% sucrose	7.2
1/10 fresh β-mercaptoethanol in T0.1E	0.187
10 mM dNTPs	1
10 × Reaction buffer	2
Dilution buffer	3.49
Taq polymerase (added last)	0.125

3. Vortex the master mix.
4. Add 1 μl of primers at 100 ng μl^{-1} of each primer and 5 μl of diluted TPEA PCR product to the master mix in the PCR tube/microtitre plate. It is always advisable to include in any experiment a positive control (e.g. genomic DNA) and a negative control (no cDNA) for each primer pair.
5. If using microtitre plate carefully place a rubber mat and gently but firmly press the hot-lid mat into place.
6. Thermocycle. We use thermocyclers from MJ Research which we have found to give even amplification across the block and possess heated lids. Our standard PCR programme following TPEA PCR is: 92 °C, 2 min; 92 °C, 30 s (× 40 cycles); 60 °C, 90 s (× 40 cycles); 72 °C, 1 min (× 40 cycles); 72 °C, 10 min.
7. Load PCR product directly into gels 2.5% agarose gels (no loading buffer required) in 1 × TBE (45 mM Tris-Borate 1 mM EDTA pH 8.0), both containing ethidium bromide, at 200 mA for 40 min. View under ultraviolet.

5 Interpretation of results

The importance of controls cannot be overstated. To test the efficacy of sample pretreatment (harvesting, reverse transcription, and amplification) we have found it prudent to test all single cell material by assaying for the presence of at least six cDNAs of abundantly expressed 'housekeeping' genes, e.g. ribosomal proteins, actin, G6PDH, etc. Only if these are present do we use the sample to investigate genes of interest. It is also highly desirable to assay for markers of cell type. For striatal cholinergic cells we have used choline acetyltransferase as a positive control. Proenkephalin and protachykinin serve as controls to detect contamination from the brain slice, as these two genes are expressed in most striatal neurons (>90%) at high abundance, but not cholinergic cells. In our analysis of these cells, of those that were positive for the housekeeping genes and choline acetyltransferase, PCR failed to detect proenkephalin and protachykinin transcripts in more than 95% of the cells studied ($n = 36$). As nuclei were not harvested from the cells, contamination by nuclear DNA and RNA was

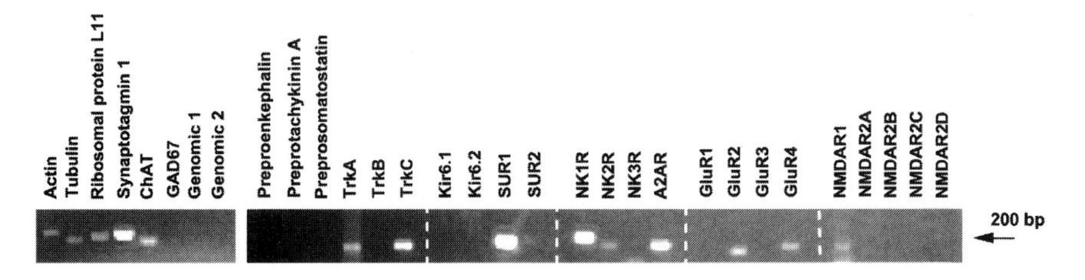

Figure 3 Analysis of gene expression in a single rat striatal cholinergic neuron. Approximately 40% of the somatic cytoplasm was harvested and subjected to reverse transcription followed by TPEA PCR. Note that two sets of primers selective for genomic polymorphic repeat sequences failed to amplify their target sequences, as did primers for genes expressed in non-cholinergic striatal neurons (GAD67, proenkepahlin, protachykinin, and prosomatostatin).

unlikely and indeed primers made to two polymorphic repeat sequences capable of amplifying genomic DNA consistently failed to amplify from TPEA PCR product generated from these cells. Other controls might include performing gene-specific PCR on cellular material that has not undergone reverse transcription or TPEA PCR. An example of the data obtained from cholinergic neurons of the rat striatum is shown in *Figure 3*.

One of the biggest problems with single cell gene expression data, whether collected by this or another approach, is the correct interpretation of negative results; either where all cells tested are negative for a particular transcript, or cases where an mRNA species is detected only in a fraction of the cells. A finding such as this could be for a number of reasons and must be interpreted with caution. It could be that the gene truly is not expressed in the cell type or expressed only in a subpopulation of the cells that had been mistakenly assumed to be homogeneous. It could, however, be that the mRNA is expressed in all the cells, but at levels below or close to the limit of detection. One way that can be used to check this is to use a greater amount of the TPEA PCR product (e.g. 6% instead of the usual 2.5%) as a template for the gene-specific assay. It there is an increase in the detection rate then it is highly likely the observation is due to a problem with the sensitivity of the assay. If not, then it is more likely that the gene is indeed expressed only in a subpopulation of cells. We have found that the more genes and cells you analyse, the greater the confidence gained in such findings.

The success of this technique can be judged by the detection of mRNAs previously shown to be expressed in these neurons. Of the genes we have studied, all those that had been shown previously to be expressed in striatal cholinergic cells by other techniques, primarily by *in situ* hybridization and immunohistochemistry, have been found to be present using this approach. Indeed if a gene is expressed, it is also potentially possible to determine the relative abundance of an mRNA in two different cell populations by determining the amount of preamplified product required in order for it to be detected (see ref. 10). With the correct controls in place, if the presence of a transcript is consistently indicated by this approach, we believe that it can be taken as strong evidence that it is present, and good evidence that the protein is likely to be present. However, additional evidence needs to be gained to support the presence of the protein, as although the expression of a gene usually results in the protein it encodes being produced, there are numerous examples where this does not hold true.

TPEA PCR provides a means to amplify of small quantities of RNA for subsequent analysis of the expression of a diverse range of genes. We have found the methodology described here to be robust and relatively simple to perform. The amplification efficiency of the current protocol is, however, more limited than theoretically possible with this approach. We are therefore currently investigating ways to improve it, with the hope of eventually being able to analyse far more genes with even greater sensitivity than possible at this time. The potential of combining the approach with other detection systems such as DNA microarrays is also highly attractive.

References

1. Lambolez, B., Audinat, E., Bochet, P., Crepel, F., and Rossier, J. (1992). *Neuron*, **9,** 247–58.
2. Sucher, N. J. and Deitcher, D. L. (1995). *Neuron*, **14,** 1095–100.
3. Yan, Z. and Surmeier, D. J. (1997). *Neuron*, **19,** 1115–26.
4. Bochet, P., Audinat, E., Lambolez, B., Crepel, F., Rossier, J., Iino, M., Tsuzuki, K., and Ozawa, S. (1994). *Neuron*, **12,** 383–8.
5. Massengill, J. L., Smith, M. A., Son, D. I., and O'Dowd, D. K. (1997). *J. Neurosci.*, **17,** 3136–47.
6. Ghasemzadeh, M. B., Sharma, S., Surmeier, D. J., Eberwine, J. H., and Chesselet, M.-F. (1996). *Mol. Pharmacol.*, **49,** 852–9.
7. Dixon, A. K., Richardson, P. J., Lee, K., Carter, N. P., and Freeman, T. C. (1998). *Nucleic Acid Res.*, **26,** 4426–31.
8. Jena, P. K., Liu, A. H., Smith, D. S., and Wysocki, L. J. (1996). *J. Immunol. Methods*, **190,** 199–213.
9. Stuart, G. J., Dodt, H.-U., and Sakmann, B. (1993). *Pflugers Arch.*, **423,** 511–18.
10. Surmeier, D. J., Song, W. J., and Yan, Z (1996). *J. Neurosci.*, **16,** 6579–91.

Chapter 9
Analysis of gene expression by two-dimensional gel electrophoresis

A. M. Tolkovsky
Department of Biochemistry, University of Cambridge, Cambridge, UK

1 Introduction

Two-dimensional (2-D) polyacrylamide gel electrophoresis using isoelectric focusing (IEF) in the first dimension separates proteins by charge and size. Although simple in concept, the technique has been under-used for three main reasons: (i) the difficulty in obtaining highly reproducible patterns of isoelectric focusing in the first (pH) dimension using tube gels and soluble carrier ampholytes; (ii) the lack of good analytical tools for analysis and assignation of differential patterns of gene expression; and (iii) inability to obtain sequencing information from other than the most abundant proteins. In the last few years, however, the practical and analytical tools have changed to such an extent that analysis of the proteome by 2-D gels can (and should) now be considered an accessible technique.

First, the introduction of immobilized pH gradients (IPG) for the first dimension (which can be purchased commercially or tailor-made to any specification) (1, 2 and references therein) allows separations of much higher resolution and reproducibility than before, the main limit to the number of polypeptides that can be detected being the range and shape of the pH gradient in the first dimension and the sizes of the resolving gels. As a result, it is now possible to detect many of the post-translational modifications, which alter the migration of proteins in the pH or size dimension (such as phosphorylation, acetylation, glycosylation) in addition to altered patterns of protein expression. Analytical user-friendly programs that use reliable algorithms for computer-based spot recognition and assignment can be purchased. Moreover, new micro-methods in sequencing and mass spectrometry (3, 4) can determine the mass of tiny amounts of peptides and sequence small peptides with such accuracy that the sequence can now be assigned to picomoles (ng) of silver-stained proteins which can be punched out, digested, and analysed directly from 2-D gels.

With the completion of the Human Genome Project, an inevitable switch in emphasis from the genome to the proteome is looming which will probably bring in its wake a huge surge in new technology. Currently, there are many variations on the theme of how to run the IEF portion of 2-D gels. In this chapter, the protocols for running 2-D gels using commercially prepared IPGs in the first dimension are described. Further useful and detailed practical information can be found in several new books (5, 6), including the 3rd edition of *Gel electrophoresis of proteins* in this series, and in special issues of the journal *Electrophoresis* .There is also much information available on the internet (e.g. http://www.expasy.ch). The main emphasis here is on the methods we have used to obtain a reproducible pattern of 2-D gels that enable multiple gels to be compared.

2 Basic methods for running 2-D gels

2.1 Sample preparation

The quality of a 2-D gel separation depends almost entirely on the quality of the sample. A good sample for 2-D separation must: (i) retain its native characteristics as much as possible, (ii) contain no precipitate, and (iii) be as enriched as possible with the proteins being analysed.

(1) To retain native characteristics, urea is used as the main denaturant with other mild detergents aiding protein extraction and solubilization. Urea comes out of solution easily at low temperatures so care should be taken that the urea is in solution. Urea solutions containing proteins should not be heated above 30 °C because this causes carbamylation of proteins at multiple sites causing a single polypeptide to appear as a series of spots in the first dimension. In fact carbamylated markers of different molecular weights can be purchased (Amersham Pharmacia Biotech) as indicators for the pH gradient. Swiftness of sample preparation is of essence to avoid degradation. Lysis buffers also contain a lot of dithiothreitol (DTT) to ensure complete reduction of S–S bonds (and prevent horizontal streaking in the second dimension) and strong inhibitors of proteolysis must be used.

(2) Samples must be completely soluble, otherwise precipitates form in the first dimension which clog the gel during transfer to the second dimension and cause streaks and blotches to form in the second dimension gel. A thorough centrifugation after lysis is therefore imperative.

(3) Urea-based lysis buffers will not extract some proteins as efficiently as sodium dodecyl sulphate (SDS). It is best therefore to grind solid tissues before lysis. For membrane proteins in particular, modifications have been introduced in which samples (2 mg ml^{-1} protein) are first extracted into 4% SDS solution (containing 100 mM Tris-HCl pH 8.5, 50 mM DTT and 20% glycerol, 2 mM phenyl sulphonyl methylfluoride (PMSF) heated at 80 °C for 3 min, spun at 12 000 g for 10 min as for normal SDS-polyacrylamide gel electrophoresis (SDS–PAGE). SDS is then removed by precipitation of proteins into 10 volumes of acetone (−20 °C) for 20 min, spun again at 15 000 g for 30 min from which they are resolubilized into the first dimension lysis buffer for 1 h after com-

plete evaporation of acetone using a vacuum pump. Again, often proteins are difficult to solubilize from acetone precipitates. We have found that use of ethanol (10:1, v/v) instead of acetone for precipitation often cures this problem but not all proteins will be precipitated by ethanol.

Protocol 1

Sample preparation

Equipment and reagents

- Phosphate-buffered saline (PBS) (150 mM NaCl, 5 mM KCl, 10 mM Na-phosphate buffer pH 7.4)
- Lysis buffer (9 M urea, 2%[a] (w/v) CHAPS (3-[3-cholamido-propyl)dimethylammonio]-1-propane sulphonate (Sigma), 0.5%[a] (v/v) Triton X-100, 20 mM DTT, 1.2% Pharmalytes pH range 3–10 (Amersham Pharmacia Biotech), 8 mM PMSF[a]
- Liquid nitrogen-cooled mortar and pestle (for grinding tissues)
- Sonicator (for grinding tissues)
- Quick method for protein determination (such as Bradford reagent)

Method

1. Wash tissue quickly in ice-cold PBS, grind in cooled mortar with pestle and suspend immediately in lysis buffer. Sonicate mixture for a further minute on ice.
2. For cultured cells, wash plates (if cells are adherent) thoroughly with ice-cold PBS, place dish on ice bucket at an angle to drain all excess PBS for 1–2 min, warm plate to room temperature and add lysis buffer. Collect sample by scraping with a rubber policeman (can be the plunger of a 1–5 ml syringe). For cells in suspension, follow the same wash protocol, except that washes should be done by centrifugation.
3. Determine protein concentration; a concentration of 2–5 mg ml^{-1} is best for analytical detection by silver staining, although up to 10 mg ml^{-1} samples have been applied to IPG strips with good results.
4. Spin sample at 15 000 g for 10 min to remove all particulate material.
5. Save supernatant sample in aliquots (of 100 μl) at −80 °C until use.

[a] PMSF has a half-life of about 30 min in water. A stock of 100 mM can be prepared in isopropanol and dispensed into lysis buffer immediately before use. AEBSF (Calbiochem) is a more stable but less reactive alternative.

2.2 Rehydration of IPG strips

Rehydration of IPG strips can be performed separately from and prior to sample application using a commercial set-up or an easily assembled home-made device. Alternatively, it is possible to swell the strip with the solution that already contains the sample (rehydration loading). The latter is a neat way of ensuring smooth impregnation of the gel with the sample without getting deposits that can occur when the sample is applied in one spot (at the anode or cathode) (Amersham Pharmacia Biotech now sells a system in which rehydration loading followed by IEF occur automatically; for evaluation of its use see ref. 7). We have used a sample volume of 500 μl for an 18 cm × 0.5 cm strip (slightly in excess of calculated volume of 450 μl) using an nuclear magnetic resonance (NMR) glass tube, which has a lid as our swelling cassette with excellent results. However, it is possible that some proteins in the sample will not enter the gel easily during passive swelling so if your sample is scarce, it may be advisable to use the more conventional route of loading the sample on to a pre-swelled gel as loading is then performed under voltage and so increases the capacity of the gel for sample entry.

Protocol 2

Rehydration of IPG strips

Equipment and reagents

- IPG Strips (Amersham Pharmacia Biotech); these strips come in various pH ranges, gradient shapes, and sizes. They are cast on a strong plastic backing, and are covered with a thinner plastic protective sheet. Keep flat in −20 °C until use. Fully swelled, they should be 0.5 mm thick
- Rehydration solution (8 M urea, 2% CHAPS, 2% Pharmalytes pH 3–10, 10 mM DTT, a few grains of bromophenol blue)
- Silanizing solution (Sigmacote, Sigma) should be used in a fume cupboard
- Rehydration cassette (two home-made glass plates separated by 0.7 mm water-proof U-shaped PVC gasket with spring clips to hold assembled plates together can be used). Glass plates should be cleaned thoroughly before use wiped so that no specs of dust or dirt adhere to the plate surfaces. Gasket should also be cleaned thoroughly to remove any protein contaminants

Methods

A Rehydration in a cassette

1. Put one of the two glass plates in a fume cupboard. Use a plastic dropper to place a few drops of Sigmacote on the plate and use a tissue to spread it lightly over the glass surface. Allow the solution to evaporate. Repeat process three times. Wash with 70% ethanol to remove residue and let dry. Solution should be reapplied every few months.
2. Take non-silanized glass plate, lay it flat, place U-shaped gasket, and moisten glass surface lightly with distilled water.
3. Remove protective (thin) plastic off strip and lay strips flat with plastic backing side down on moistened glass surface so that the moisture traps them securely in place.
4. Place second (silanized) glass plate on top, silanized face down, clip both plates together and stand cassette vertically.

Protocol 2 continued

5 Pour rehydration buffer into cassette ensuring that the entire surface of the strip is covered; cover with cling film to prevent evaporation and leave overnight to swell.

B Rehydration of sample direct into the gel—rehydration loading

1 Calculate the precise volume of solution required for swelling an individual strip.

2 Put the sample (in lysis buffer) in a shallow purpose-made well or use an NMR tube and gently place strip with free surface facing downwards into the sample.

3 Cover the well to avoid evaporation and gently shake on a side-by-side rotator overnight.

2.3 Running the first dimension

The method outlined below makes use of a programmable power supply. This is useful because it is advisable to begin with a low voltage setting (100–300 V) to facilitate sample entry before switching to the final voltage. However, the main principle is to obtain voltages of 3.5–5 kV in order to accumulate between 30 and 100 kVh total voltage × time for good steady-state separations to be completed within 24 h. Any power supply that can give such output is sufficient. If the strips were swelled without the sample, the sample must then be applied separately and become absorbed into the gel. If the sample is in a small volume, then it can be added directly on to the rehydrated strip. We have added dilute labelled samples of sympathetic neuron proteins in small volumes (18 μl) directly on to the rehydrated strips (at the anode end) with good results. Alternatively, the Multiphor system comes with sample cups. These can be tricky to operate because they must touch the gel surface so as to form a closed chamber without gouging into it thus causing discontinuity, so care is required. The biggest problem with a sample added at one gel end in a small volume is that proteins can precipitate at the site of deposition. To offset this problem, one should try loading sample at the anode and at cathode ends. It is also important to begin with a

Table 1 Typical first dimension IEF running voltages for 18 cm IPG 3–10 nonlinear gradient strips. We used a total time in volt-hours (Vh) of 66.5 KVh. This value depends on length of strip, range and type of IPG gradient and amount of protein. It is important not to underfocus (streaks) but also not to overfocus (urea crystals because of water leaching out of gel strips, fuzzy protein spots). See Discussion in ref. 2 for further comments

Volts	mA	Watts	Time	Comments
0–150	1	5	1 min	To reach volatge
150	1	5	30 min	Aid sample entry
150–300	1	5	5 min	Reset voltage
300	1	5	1 h	Further sample entry
300–3500	5	5	3 h	Reset volatge
3500	5	5	15–20 h	Separation to equilibrium

gentle application of 100–300 V for 1–3 h to facilitate sample entry. Finally, running temperature should be about 20 °C. At lower temperatures, urea can crystallize out, causing water to follow, ultimately distorting the separation pattern and not allowing complete steady-state conditions.

Protocol 3

Isoelectric focusing

Equipment and reagents

- Multiphor II horizontal electrophoresis apparatus with Immobiline Drystrip tray (which can be reused) (Amersham Pharmacia Biotech)
- Power supply that yields 3.5 kV (Amersham Pharmacia Biotech is programmable)
- 3MM Whatman paper for making wicks
- Chilling water bath that can maintain a temperature of 15 °C
- Sample cups (Amersham Pharmacia Biotech)
- Paraffin oil (Dow Corning 100/10cS or Sigma Mineral oil M-5904)

Method

1. Pour off rehydration solution and open swelling cassette.
2. Blot each strip between two wetted filter papers that are just moist
3. Align strips in the Immobiline Drystrip tray such that the pointed end is facing the anode (+) electrode and the blunt end is facing the cathode (−) end. (Strips can be identified by scratching numbers into the plastic backing at the tip.)
4. For wicks, cut two strips of 3MM Whatman paper about 0.5–7 cm wide and moisten with de-ionized water[a] (wicks should be moist—not wet). Place one strip across the (+) end and one strip across the (−) end of the gels.
5. Lay appropriate electrodes on wicks.
6. Place sample cups according to the manufacturer's instructions. Be sure to centre the cups over the strips and press them down slowly so that they touch the surface of the swelled strips firmly but do not gouge out the gel. This is a bit tricky.
7. Optional: Cover sample cups and strips with light paraffin oil. Note: This step is useful because oil will leak through cups which are not adhering to the gel properly and so indicate a problem. If this happens, press the cups down again gently and make sure there are no leaks. Oil seepage doesn't affect the end result. However, oil is not absolutely necessary to get good separations.
8. Add samples to cups under the oil layer. It is best to have a dilute sample in 100 μl rather than a concentrated sample in a small volume because this way precipitation of the sample at the point of entry is avoided. Do not use less than 20 μl per cup.
9. Connect circulating water bath set at 15–20 °C.

Protocol 3 continued

10 Connect power pack and run. A typical run regimen, and some comments on the number of volt-hours required can be found in *Table 1*. Note: if there is a power failure for a period of time up to overnight do not panic. The system is usually robust enough such that power can be reconnected and the run resumed without major problems.

11 Optional: Samples in <20 μl can be added directly to the strip (we added the samples at the anode end). In this case, begin the IEF and add paraffin oil only after the samples have entered the gel.

12 At end of the run, disconnect power supply, remove IPG strips, and either process them directly for second dimension or wrap each strip carefully without damaging surface in aluminium foil and store flat at −20 or −80 °C.

[a] Some protocols recommend use of 10 mM glutamate for anodal wick and 10 mM lysine for the cathodal wick; water works just as well.

2.4 Equilibration of IEF strip before running second dimension

To obtain reproducible multiple 2-D gels, the same set of proteins in each sample must be transferred from the first to the second dimension SDS-polyacrylamide gel. Ideally none of the sample should be left in the IEF strips. Various protocols have been described that: (i) facilitate SDS binding prior to loading strips on to the second dimension gel; (ii) lower concentration of the urea, which can obstruct transfer, especially if it is crystalline; and (iii) prevent oxidation of the thiol groups and formation of S–S bonds (second dimension gel may retain some ammonium persulphate (APS) which will also oxidize proteins so blocking SH groups with iodoacetamide will prevent this. If however, proteins are to be sequenced, covalent modification may not be a good idea). It should be noted, however, that some proteins will invariably leach out into the equilibration buffer so that a balance must be struck between the time required for proper equilibration and the possible loss incurred.

Protocol 4

Equilibration of strips before transfer to SDS–PAGE

Equipment and reagents

- Large Petri dish or Perspex block containing wells of 0.8 cm × 20 cm × 1 cm
- Equilibration Buffer 1: 50 mM Tris–HCl buffer pH 6.8, 6 M urea, 30% (w/v) glycerol, 1% (w/v) SDS, 2.5 mg ml^{-1} DTT
- Equilibration Buffer 2: 50 mM Tris–HCl buffer pH 6.8, 6 M urea, 30% (w/v) glycerol, 1% (w/v) SDS, 45 mg ml^{-1} iodoacetamide, 30 μ g ml^{-1} bromophenol blue

Protocol 4 continued

Method

1. Submerge strips in Equilibration Buffer 1 for 10 min at room temperature.
2. Wipe lightly on tissue to remove excess fluid
3. Submerge strips in Equilibration Buffer 2 for 5 min at room temperature.
4. Optional: Cut off 1–1.5 cm from either end as usually these areas were covered by the wicks and no focusing would have occurred around these regions. For an 18 cm strip, this leaves a strip length of 15 cm, which will fit most medium size PAGE systems.

Protocol 5

Casting the second dimension gel

Equipment and reagents

- Any PAGE system (thickness 0.75 mm) that will support the length of the IPG strips. A square format gel seems to work best
- Resolving gel: acrylamide/bisacrylamide (a mix of 10% Total acrylamide + bisacrylamide, 2.67% bisacrylamide relative to total will show an actin spot, 43 kDa, about halfway down the gel), 375 mM Tris-HCl pH 8.8, 0.1% SDS (w/v), 0.05% ammonium persulphate (w/v), 0.05% (v/v) N′N′N′N′ tetramethylethylenediamine
- 1% (w/v) low melting point agarose (Sigma) in 150 mM Bis-Tris, 0.1 M HCl, 0.2% SDS (w/v)
- Small paper strips (0.5 × 1.5 cm) as identifying labels. Use HB pencil to mark each strip (date, number, etc.)
- Fixing solutions (I) 50% ethanol, 12.5% acetic acid (1 litre for six gels, 20 cm × 20 cm) for drying down directly or as a prelude to silver staining

Method

1. Place a label in a bottom corner of the glass cassette and cast the resolving gel (regular paper does not float but some filter papers may float depending on gel density). After polymerization wash surface gently with distilled water.
2. Heat the agarose solution to melt, cool to about 60 °C and pour on to resolving gels.
3. Immediately immerse the IPG strips through the agarose until they just touch the surface of the gel.
4. Run sample as you would normally for a 1-D gel. The faster the run, the more focused the spots. However, spot yawning may occur if the temperature rises.
5. At the end of the run, remove each gel carefully.

6a If gels are to be dried down without staining, or stained with silver nitrate, gel can be immersed in fixing solution 1 h to overnight at room temperature with gentle agitation.

6b If gels are to be blotted transfer on to PVDF membrane without fixing.

6c Gels may be fixed and stained in Coomassie Brilliant blue or colloidal solutions of Coomassie (see ref. 12 for these and other dyes).

2.5 Running the second dimension vertical SDS–PAGE

A prerequisite for successful comparisons between several 2-D gels is to be able to manufacture perfect replicas. The best means of achieving such replicas is to run the second dimension gels in the same gel tank under constant temperature. Anderson and Anderson were the first to describe a horizontal gel system, which is known as the Isodalt system (8). We have described the design and manufacture of our own temperature-controlled second dimension tank which runs six vertical 20 × 20 cm gels simultaneously (9). For SDS–PAGE we have used the standard Laemmli buffer system and protocol (10, see also ref. 11 for up-to-date review of various 1-D systems currently available) so this will not be elaborated upon here (for an evaluation of horizontal versus vertical second dimension gels, see ref. 2). For vertical gels, the bigger the size of the plates, the higher the resolution. The second dimension gel can be cast with or without a stacking gel as the IPG strip acts as a surrogate stacking facility. Many researchers recommend loading strips through using an overlay of 1% (w/v) low melting agarose heated to about 60 °C. Whichever system is used, a most important aspect is to load the strip into the gel plates so that it remains flat. This is especially important when protein patterns are to be compared between samples that have been labelled with weakly radioactive isotopes such as [^{35}S]-methionine as differences in the thickness of the dried gels will cause the signal to be quenched unevenly. We found that the best method was to tease the strip such that the plastic backing adhered to the back glass plate.

2.7 Drying the gels

Drying large slab gels can cause large distortions that vary between gels due to uneven shrinkage. To avoid these problems as much as possible, it is best to avoid drying gels rapidly at high temperatures. We devised a slow drying protocol that is outlined below.

Protocol 6

Slow drying of 20 × 20 cm slab polyacrylamide gels

Equipment and reagents

- Wash solution: 30% (v/v) ethanol
- Glycerol impregnation solution: 30% ethanol containing 4% (v/v) glycerol (to provide gel with some flexibility)
- Silylated glass plates (22 × 22 cm) (procedure described in *Protocol 2*.)
- Squares of Cellophane about 5 cm larger than the glass plate on which the gel will be dried) (BioGelWrap, Biodesign Inc.)
- Squares of Whatman 3MM paper (about 5 cm bigger than the glass plates)
- Photocopy film

Method

1 Wash gels in 30% ethanol for at least 1 h (to remove acetic acid fixative).

Protocol 6 continued

2 Impregnate gels with Glycerol solution 2×30 min.

3 Place one gel on to the silylated surface of the glass plate (written surface of paper label facing glass plate).

4 Immerse cellophane in the glycerol impregnation solution until it is completely wet.

5 Place sheet of cellophane on top of the gel, remove all air bubbles and wrap the excess cellophane around the plates such that the surface is smooth and taut.

6 Place the cellophane-wrapped glass plate (gel facing up) firmly on top of one or two Whatman 3MM sheets and allow to dry over night at room temperature or for a few hours at 35 degrees.

7 Optional: Spray photocopy film with Spraymount (3M) and glue to the cellophane side of the dried gels.

8 Remove glass plate, trim excess cellophane/photocopy sheet. Store dried gels in A4 plastic pockets.

3 Special techniques for detection of differential gene expression

3.1 Metabolic labelling

Often it may be desired to obtain a differential fingerprint of proteins induced within a short time after a treatment (for example, with a growth factor). In such cases, *in vitro*, it is possible to capture the production of these proteins by metabolic labelling. Metabolic labelling with [^{35}S]methionine is compatible with matrix-assisted laser desorption/ionization time of flight mass spectrometry and peptide sequencing using electrospray ionization methods or N-terminal sequencing of proteins from PVDF blots. Administration of carrier-free [^{35}S]methionine will give maximum specific activity of proteins and thus increase the sensitivity of detection but may not yield enough protein for analysis by microsequencing. High sensitivity may be combined with higher yields by diluting carrier-free labelling medium with a methionine containing medium, or spiking a cold protein extract with the 'hot' extract. There are two main products for labelling with [^{35}S]-amino acids, one in which methionine alone is used in a highly purified form, and one which contains a mixture of [^{35}S]-labelled methionine and cysteine (Translabel, ICN). The latter is derived from a hydrolysate of bacteria fed with ^{35}S sulphur, and may contain some trichloroacetic acid (TCA)-precipitable material. This may add background when a TCA-precipitate obtained from the labelled material is used to calculate total incorporation.

Protocol 7

Metabolic labelling

Equipment and reagents

- L-[^{35}S]methionine/cysteine >1000 Ci mmol^{-1} (Amersham Pharmacia Biotech SJ1015)
- Methionine-free medium (any commercial supplier)
- Glass fibre filters (Whatman GF/C)
- BSA solution (2 mg ml^{-1} bovine serum albumin Fraction V (Sigma))
- 25% Trichloroacetic acid (w/v) in water (take care: solution very corrosive)
- Scintillation fluid (Optiphase Hisafe FSA)

Method

1. Incubate tissue/cells for 10 min in methionine-free medium to lower endogenous methionine levels.
2. Dilute 500 μCi [^{35}S] into 1 ml of methionine-free medium and add back 1/20 volume of full medium (to avoid reduction in rates of protein synthesis due to limiting concentrations of amino acids).
3. Label cells for the desired period.
4. Wash our excess label using a medium containing the normal medium containing methionine.
5. Wash cells in ice-cold PBS and process as described in *Protocol 1*.
6. Estimate amount of [^{35}S] incorporated by adding a known volume (1–5% of total) to 250 μl of BSA solution; then add 250 μl 25% TCA and leave at 4 °C for at least 10 min.
7. Decant on to a glass fibre filter prewetted with 10% TCA solution and maintained under suction, wash filter three rimes with 10 ml ice-cold 10% TCA, dry filter, and count using an aqueous grade scintillation fluid.

3.2 Making a calibration strip for quantitation of [^{35}S]methionine-labelled spots

If a set of 2-D gels are used to compare between several samples of [^{35}S]methionine-labelled proteins where differences in intensities are used to indicate different levels of expression, it is imperative to prepare calibration strips with known amounts of radioactivity to offset any differences in exposure time/quality of film or storage phosphoimaging screens. Hence, the sum total intensity of all the spots can be scaled to that of the calibration strip before spot quantitation. Methionine has a half-life of 84 days so [^{14}C]-labelled proteins (half-life ~5000 years) are used instead. To obtain [^{14}C]-labelled proteins, we labelled a set of proteins with [^{14}C]N-ethylmaleimide (NEC454, 20–40 mCi mmol^{-1}, New England Nuclear) (proteins were a mix containing 1 mg ml^{-1} of cytochrome *c*, RNase A, creatine kinase, actin, BSA, and β-galactosidase) (14). However, the most inexpensive [^{14}C]-labelled protein product from any commercial supplier can be used instead as long as one can achieve 2–4 × 10^6 c.p.m. ml^{-1}.

Protocol 8

Preparation of a calibration strip made of [^{14}C]-labelled proteins

Equipment and reagents

- Minigel apparatus
- Capable of yielding 2–4 × 10^6 c.p.m. ml^{-1}
- 10% TEMED (Sigma)
- 8 ml of incomplete gel mixture (acrylamide/bisacrylamide 10%T/2.67% C in water, 0.075% APS) but no TEMED

Method

1. Prepare 1 ml aliquots of the incomplete gel mixture.
2. To each of four aliquots add [^{14}C]-labelled protein(s) to obtain a range of gel solutions containing a spread of values from approximately ~20 000 to 2 000 000 c.p.m. ml^{-1}.
3. Count a 10 μl sample from each solution to measure the precise radioactivity in the sample (in c.p.m. ml^{-1}).
4. Cast sequential layers as follows, adding TEMED (0.03% final concentration) only to the layer being cast. Be sure to let each layer polymerize before adding the next layer (see *Figure 1*).
5. Dry the gel.
6. Cut out strips at 90 °C to the direction of the layers.
7. Use one calibration strip for each gel being exposed.

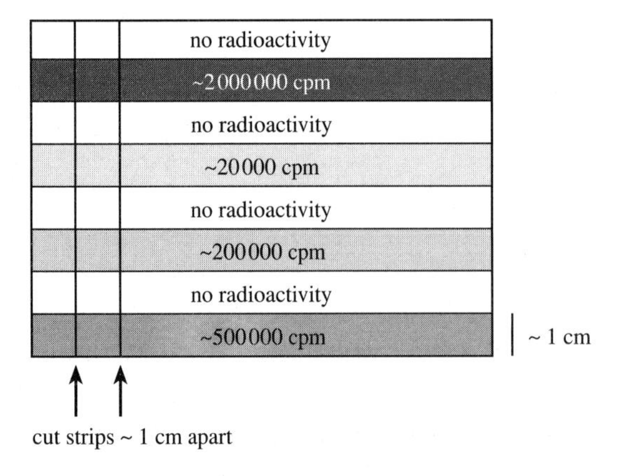

Figure 1 Preparation of calibration strips.

3.3 Silver staining that is compatible with matrix-assisted laser desorption/ionization mass spectrometry and electrospray mass spectrometry

The most recent development in obtaining sequence information from subpicomole (more realistically picomole) amounts of protein cut out from polyacrylamide gels is analysis by tandem mass spectrometry (3, 4; see ref. 13 for detailed protocols). Other protocols can be obtained from Protana (http://www.protana.com/). Silver-stained proteins obtained for this application should not be fixed with glutaraldehyde prior to silver staining. A protocol that is compatible with MS/MS sequencing is given below. These gels shrink during fixing due to high methanol content so we use ethanol fixative (as in *Protocol 8*) for preparing silver-stained gels for comparative pattern analysis. For least background, make up all solutions fresh with pure chemicals. It cannot be emphasized enough that hands (and hair) shed keratin continuously. Keratin contamination is a real nuisance so gloves must be used during all procedures. For the preparation of blots on PVDF membranes for Edman sequencing consult http://www.bio.cam.ac.uk/proj/adr/PNAC/blotguide.html.

Protocol 9

Silver staining for MS/MS

Reagents

- Fixing solution 1: 50% methanol, 5% acetic acid
- Fixing solution 2: 50% methanol
- Sensitizing solution: 0.02% thiosulphate
- Staining solution: 0.1% silver nitrate in water (chill to 4 °C)
- De-ionized water
- Developer: 2% sodium carbonate containing 0.04% formaldehyde
- Stopping and storage solution: 5% and 1% acetic acid

Method

1. Fix gel in Fixing solution 1 for 20 min.
2. Wash out acetic acid in Fixing solution 2 for 10 min.
3. Wash three times with deionized water to remove all traces of acid.
4. Sensitize proteins with thiosulphate for about 1 min.
5. Rinse 2 × 5 min with water.
6. Submerge gel in silver nitrate solution at 4 °C for 20 min.
7. Transfer gel to a new clean dish that hasn't been in contact with silver nitrate and rinse gel two times 1 min with water.
8. Add developer under constant supervision, making sure that developer is changed promptly if a yellowish colour develops.

Protocol 9 continued

9 When spots are observed, stop development by replacing developer with 5% acetic acid.

10 Wash gel once in stopping solution and store gel in 1% acetic acid.

4 Pattern analysis and spot assignment

Although it is possible to spot some differences between high-resolution gels by eye it is impossible to store the data or quantify patterns of gene expression from multiple high-resolution gels. Indeed, due to the inherent variability of the 2-D gel technique, a reliable pattern can only be obtained by running several samples of the same, or cohort, sample. It is therefore imperative to have a system that will scan/digitize the images, and then identify the spots by assigning each spot with a specific address. The next task is to compare the intensity of the spots between samples such that changes in gene expression can be defined. Then, ideally, proteins of interest can be excised and sequenced. Hanash (15) lists several Web sites for imaging software developers. Some of these require a Unix SUN system, which is not available in every lab but may provide a service, or other information (see *Table 2*). If, however, there are only two developmental states or treatments of interest, it maybe possible to circumvent running several gels, by using the approach outlined by Uhlu *et al.* (16) named difference gel electrophoresis. This approach entails covalent labelling of proteins from two samples with different fluorophores (in this case, Cy3- and Cy5-based *N*-hyxdroxy-succinimide esters). Samples 1 and 2 were then combined and run on a single 2-D gel. The gel was scanned with two wavelengths, whose subtraction from each other yielded spots for proteins that were differentially expressed. Covalent modification of proteins using this method is not perfect because many proteins may not be labelled, and if spot assignment is desired, the modification could alter the IEF/second dimension pattern.

Table 2 Suppliers of software and other useful services

Web site	Analytical system	Other services
http://bioimage.com	Visage™ (Unix)	Protocols
http://www.expasy.ch	Melanie II (Mac/PC/Unix)	Protocols[a]
http://www.phoretix.com	Phoretix '98 (PC)	Links to other info
http://scanalytic.com	Gelab II+ (Windows)	Little info
http://members.aol.com/aabsoft	AAB (Mac, Windows, DOS)	Low priced

[a] Some important and useful changes to standard protocols are suggested.

5 Some basic troubleshooting

There is no doubt that consistency of gel patterns improve with practice. It is always a good idea to run duplicate gels from each sample/condition to control for run variability. To improve gel quality the following points may be considered:

(1) To avoid local precipitation at the site of sample application on the IPG strip it is better to load the sample in a larger volume than as a more concentrated solution in a small volume. If necessary, repeat the application and run a low voltage until all the sample is absorbed.
(2) If loading the sample at the anode end seems to be a problem, try loading at the cathode end.
(3) Larger amounts of protein require longer separation times in the first dimension.
(4) Samples containing high salt may not separate well.
(5) If streaking during IEF is a problem, try placing a piece of Whatman 3MM paper impregnated with 15 mM DTT adjacent to the cathode wick (in the direction of the anode). DTT will migrate towards the anode and hence ensure proteins remain reduced.
(6) Avoid temperature rises during electrophoresis.
(7) Do not use old reagents—use high-grade/purity chemicals.

Acknowledgements

I wish to thank Bob Amess, who developed and implemented many of the techniques described in this article while he was a postdoc in the lab. Our work was funded by the MRC and the Wellcome Trust.

References

1. Bjellqvist, B., Ek, K., Righetti, P. G., Gianazza, E., Gorg, A., Westermeier, R., and Postel, W. (1982). *J. Biochem. Biophys. Methods*, **6**, 317–39.
2. Gorg, A., Boguth, G., Posch, A., and Weiss, W. (1995). *Electrophoresis*, **16**, 1079–86.
3. Wilm, M., Shevchenko, A., Houthaeve, T., Breit, S., Schweigerer, L., Fotsis, T., and Mann, M. (1996). *Nature*, 379, 466–9.
4. Shevchenko, A., Wilm, M., and Mann, M. (1997). *J. Protein Chem.*, **16**, 481–90.
5. Westermeier, R. (1997). *Electrophoresis in practice: a guide to methods and applications of DNA and protein separations*. VCH, Weinheim.
6. Link, A. J. (ed) (1998). *2-D Proteome Analysis Protocols. Methods in Molecular Biology* Vol 112, Humana Press.
7. Gorg, A., Obermaier, B. C., Hadar, A., and Weiss, W. (1998). *Life Sci. News*, **1**, 4–67.
8. Anderson, N. G. and Anderson, N. L. (1978). *Anal. Biochem.*, **85**, 331.
9. Amess, B. and Tolkovsky, A. M. (1995). *Electrophoresis*, **16**, 1255.
10. Laemmli, U. (1970). *Nature*, **227**, 680–5.
11. Shi, Q. and Jackowski, G. (1998). In *Gel electrophoresis of proteins: a practcial approach* (3rd edn). (ed. B. D. Hames), p. 1. Oxford University Press, Oxford.
12. Merril, C. R. and Wasgarf, K. M. (1998). In *Gel electrophoresis of proteins: a practcial approach* (3rd edn). (ed. B. D. Hames), p. 53. Oxford University Press, Oxford.

13. Jensen, O. N., Shevchenko, A., and Mann, M. (1997). In *Protein structure: a practical approach* (2nd edn). (ed T. E. Creighton), p. 29. Oxford University Press, Oxford.
14. Evans, T., Rosenthal, E. T., Youngblom, J., Distel, D., and Hunt, T. (1983). *Cell*, **33**, 389.
15. Hanash, S. M. (1998). In *Gel electrophoresis of proteins: a practcial approach* (3rd edn). (ed. B. D. Hames), p. 189. Oxford University Press, Oxford.
16. Unlu, M., Morgan, M. E., and Minden, J. S. (1997). *Electrophoresis*, **18**, 2071.

Chapter 10

Proteome research: methods for protein characterization

Ian Humphery-Smith
Department of Pharmaceutical Proteomics, University of Utrecht, The Netherlands

Malcolm A. Ward
Glaxo Wellcome Research & Development, Stevenage, Hertfordshire, UK

1 Introduction

The term proteomics was first introduced into the scientific literature in 1995 to define 'the total protein complement of a genome' (1) and has become an integral part of gene-expression analysis (2). However, proteomics did not represent a new scientific discipline, but rather one searching for an identity in the Age of Genomics. Proteomics is still largely centred upon two-dimensional (2-D) gel electrophoresis for the initial separation of complex mixtures of cellular proteins. Although this separation science has been with us for some 25 years, it is only recently that the associated level of reproducibility has undergone marked improvement, while these advances have also been coupled with advances in the application of mass spectrometry to protein science. Over the last decade, mass spectrometry has become increasingly user friendly and a powerful analytical device. These tools now deliver high mass accuracy, resolution and sensitivity to a point where it is currently impurities associated with sample handling and delivery that limits the threshold of detection for this versatile analytical platform. Mass spectrometry in the protein laboratory is now the preferred option for protein characterization and can be conducted efficiently on purified proteins present at low femtomole concentrations (3). Other approaches conducted independently to the separation sciences are starting to make their appearance and fall under the heading of second generation proteomics, which utilizes array-based approaches to follow protein abundance and protein interactions (4).

Proteomics and its applications to the biomedical sciences have been reviewed (5–18), as has its potential utility to the biotechnology (19, 20) and pharmaceutical industries (21, 22). It is not possible to provide a list of all procedures currently employed in protein chemistry and of relevance to proteomics. Thus, it has been decided to provide the reader with a series of protocols able to guarantee a successful outcome to both novices and accomplished practitioners interested in characterizing a protein or proteins of interest. However, the reader is referred to a number of texts devoted to methodologies in protein chemistry (23–29). The principal objective of proteomics is to rapidly attribute either 'novel' or 'previously studied' status to a protein, so as to make an informed decision on whether or not to expend further resources on studying a particular protein. Data management and the construction of relational databases is essential to all aspects of high throughput proteomics, as it is for all aspects of genomics. The integration of image databases and analytical chemistry form the basis of much of the above, but increasingly the need for novel tools in proteomics goes far beyond. If proteomics is to become the mainstay of functional genomics, tools need to be developed that focus on manipulating both theoretical and experimentally acquired data to provide value-added insight into gene structure and function (30).

A variety of excellent texts have dealt with a practical approach to 2-D gel electrophoresis (27, 31–41) and this aspect of proteomics will not be dealt with here, apart from stressing the importance of producing good quality gels with well-focused protein spots; lacking both vertical and horizontal streaking; and situated on clear background following protein revelation. The latter is essential for the construction and comparison of representative image databases and subsequently for obtaining good results during protein characterization. 2-D gel electrophoresis remains unequalled in its ability to resolve complex mixtures of proteins. Occasionally, co-migrating spots derived from organisms with fully sequenced genomes can be defined unambiguously. However, this problem is better overcome by increasing either the real (42–46) or apparent gel size, 'proteomics contigs' (8, 47), so as to reduce the statistical likelihood of co-migration, i.e. the number of expected spots per cm^2. Sample fractionation with respect to protein complexes (48, 49), cellular organelles, molecular mass, and charge can also be employed to achieve a desirable reduction in sample complexity and thus a reduced likelihood of co-migration of protein spots. As a general rule, we have found that isoelectric focusing in the first dimension for at least 60 000–100 000 Volt/Hours (V/H) can greatly increase sample purity and the associated signal-to-noise ratio within a relatively small gel area (50). Indeed, this signal-to-noise ratio is far more important to the success of protein characterization than is the amount of protein loaded on to a gel or present in an individual band or spot. Although linked to a lesser resolving power than 2-D gel electrophoresis, 2-D high-performance liquid chromatography (HPLC) has achieved resolution of up to 600 proteins from cellular extracts and can provide useful resolution prior to the delivery of protein samples to mass spectrometry (51).

The diversity of properties possessed by individual proteins can vary dramatically and reflects directly the fantastic diversity afforded by the order, composition, and conformation of some 20 amino acid building blocks. The abundance, size, pI, hydrophobicity, and mass of the latter, all then combine to define the ease or difficulty with which proteins can be studied in a laboratory setting. This is still further complicated by co- and post-translational modifications. The objective of the present chapter is to help ensure quality results for someone being introduced to the field or an experienced scientist experiencing difficulties with one or more procedures in protein chemistry. In addition, Capillary Zone Electrophoresis (CZE), HPLC, and mass spectrometry must be optimized with respect to individual samples and the manufacturer's recommendations for use. By ensuring some initial success, it is hoped the reader will be stimulated to persist within a demanding discipline associated with ever-expanding horizons and experimental nuances.

2 Cleanliness

As the sensitivity of detection, the associated mass accuracy and resolution of analytical platforms increases, then so too does the need for increased attention being paid to the maintenance of sample purity. Although good protein resolution has also been achieved with 2-D HPLC (51), extended isoelectric focusing of complex mixtures of proteins during 2-D gel electrophoresis currently produces individual protein spots of unsurpassed purity. Indeed, the impurities have by definition migrated out of the sieving matrix or have been concentrated near to their own pI and M_r. The dilemma for the protein scientist is how best to maintain this purity during sample processing and the delivery, for example, to the mass spectrometer.

Of critical importance here is water quality. It is a good rule to use only the highest purity reverse osmosis, resin-purified water that has not been stored, i.e. for all water needs wait for freshly purified water to be delivered into sterile and dust-free containers. Some researchers also double distil the output from the former configuration. The old adage of 'garbage in equals garbage out' is never more true than during protein characterization. Good results can also be obtained with analytical grade water purchased from a commercial source. Here, however, the danger arises once the bottle of high purity sterile water is opened. The contents of these bottles must be discarded regularly (kept for no longer than 5 working days and preferably less), so as to prevent the establishment of contaminating biofilm.

2.1 Dust reduction hoods

Dust hoods can provide a very cost-effective means of reducing keratin levels in samples, as can maintenance of a cover at all times over microtitre plates during pipetting operations. Dust reduction hoods should be constructed out of solid plexiglass as deep as the laboratory bench and 70–90 cm high and about 1 m wide. Doors on the front of these structures help to reduce further air flow and dust deposition when the surface is not in use. The hoods are well employed in conjunction with sample preparation areas and the balance bench.

- Keep these shut when not in use
- Maintain a spotlessly clean working environment within the hood
- These hoods will also facilitate more accurate weighing due to reduction of air flow
- Visibly dirty Perspex should be wiped clean regularly

2.2 Filtering buffers and solutions to remove particulate matter and micro-organisms

Disposable syringe filters and bell filters are useful for the removal of living and larger particulate matter (hair) from buffers and solutions (<500 ml) to be used in sample preparation during protein characterization.

2.3 Aerosol prevention filters in micropipette tips

Aerosol prevention filters in micropipette tips were designed specifically for applications in polymerase chain reaction (PCR) technology; however, they serve to avoid liquids making their way up into the pipette head and also to prevent contamination derived from a dirty pipette head making its way back into the liquid or sample during pipette blow-out.

2.4 Laminar flow hood

Following electrophoretic separation, all manner of contaminants can affect the signal-to-noise ratio associated with a protein sample. Thus, a Class II microbiological containment cabinet can be employed to reduce the risk of acquiring commonly encountered air-borne contaminants, such as keratin, during sample preparation.

A dirty laminar flow hood or one that is employed inappropriately can be far more inefficient than working unprotected on the laboratory bench. This is because huge volumes of dirty air being sucked into the front of the cabinet or potentially circulated within it. Movement of hands or non-sterile objects above the object wishing to be protected can jeopardize sterility due to air movement from above. Thus, a clean laminar flow hood and the maintenance of a clean and unencumbered working surface (reduction of air turbulence) within the cabinet are essential to reduce the chances of contamination. For the mass spectrometry laboratory, sterility and infection control can be used to equate with keratin reduction from sources other than water, chemicals, buffers, or in-line contaminants acquired in robotic space.

Protocol 1

Set-up routine for laminar-flow hood

1. Wear a lab-coat, hair cover (bonnet) and safety glasses, while handling protein samples.
2. Remove the front cover plate and place it on the ground (standing up near to the left of the cabinet).
3. Note the time, user name and particulars in a hard-covered notebook destined to record cabinet use.
4. Turn *on* the air flow and wait a minimum of 20 min (for sterile air flow to be established).

Protocol 1 continued

5 Attach a self-adhesive rubbish bag at the side of the cabinet.

6 During step 4, set up the work space in an uncluttered manner. (Cluttered air space creates turbulence and the risk of skin, dirt, and hair contamination of your samples.) Place a pipetting tool on the working surface (leave this pipette in the cabinet at all times).

7 Work cleanly and use individually wrapped sterile pipettes. Any rubbish (pipette wrappings, etc.) are to be placed immediately in the self-adhesive bag, as mentioned in step 5 (again to avoid cluttering the work space).

Protocol 2

Shut-down routine for laminar-flow hood

1 Remove all material from the cabinet, placing all liquid containers, pipettes, etc. in a *red square* to the *right* of the cabinet (none of this material should be allowed to make its way back into the cabinet.

2 Remove *all* material from the cabinet, except the Pipettus (to be left hanging-up inside cabinet).

3 Using a paper hand-towel, wipe-down the stainless steel work surface with an ample supply of alcohol from a squirt bottle.

4 Close the waste bag, using its adhesive strip.

5 Note the exact time in the record book for cabinet use (important for the next user).

6 Turn *off* cabinet; Turn *on* the ultraviolet (UV) light and replace white front cover.

7 Wait 20–30 min prior to again using the cabinet.

3 Western blotting

The use of antibodies remains a reliable means of identifying proteins of interest within complex mixtures of proteins. However, it presumes prior knowledge of the protein of interest and must be combined with significant prior effort directed towards the production and screening of polyclonal, monoclonal, or phage-derived antibodies. This approach is not easily amenable to large-scale proteome analysis using currently available technologies due to the inherent time and cost-intensive nature of antibody production. Excellent reviews on this topic have appeared (52, 53). It is noteworthy that one is regularly surprised by the apparent lack of specificity of both polyclonal and monoclonal antibodies exposed to 2-D blots, as their specificity has been usually determined in a 1-D environment. Monoclonal antibodies attribute their specificity to the recognition of approximately five to eight amino acids. The incidence of a particular amino acid string within a given organism, cell, or tissue type ultimately depends upon

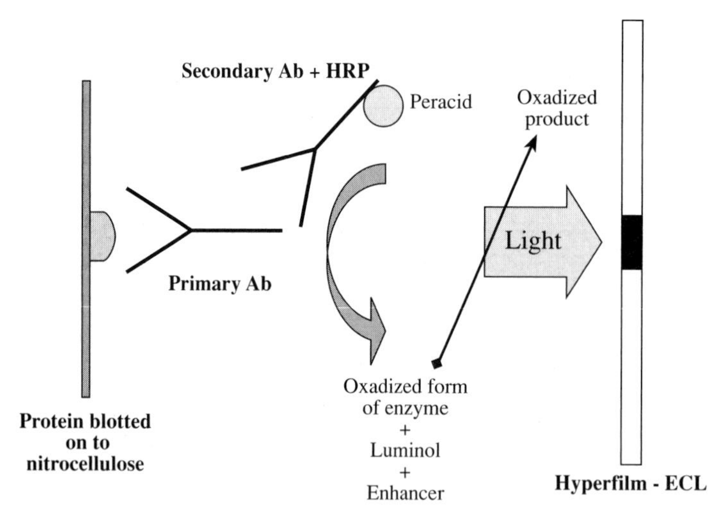

Figure 1 Schematic representation of revelation of protein spots by enhanced chemiluminescence.

the incidence of this string within the translated genome. The principles of revelation using enhanced chemiluminescence are presented in *Figure 1*.

Protocol 3

Electroblotting

Equipment and reagents

- Polyacrylamide electrophoresis gel
- Nitrocellulose Hybond-C Super (Amersham Pharmacia Biotech) membrane
- Continuous buffer solution: glycine 25.7 mM; Tris 48.0 mM; sodium dodecyl sulphate (SDS) 1.3 mM; methanol 200 ml; made up to 1 litre with distilled water

Method

1. Cut 12 pieces of clean filter paper to the exact size of your electrophoresis gel. These will ensure that current passes only through the gel and does not short circuit between the cathode and the anode.
2. The proteins present in the polyacrylamide gel used for their electrophoretic separation will undergo a semi-dry transfer to nitrocellulose Hybond-C Super (Amersham Pharmacia Biotech) membrane by creating a potential difference between two graphite plates. These plates must not be contaminated with proteins from the hands. Therefore, wear gloves.
3. Wet six pieces of filter paper by capillary action in continuous buffer solution (glycine 25.7 mM; Tris 48.0 mM; SDS 1.3 mM; methanol 200 ml; made up to 1 litre with distilled water).

Protocol 3 continued

4 Pre-wet the anode with distilled water and wipe clean with paper towel. Place the above pieces of filter paper on the anode (red/+). Pressing lightly, roll with a clean plastic pipette to remove any bubbles which might otherwise interfere with electrotransfer.

5 Place the Hybond_C Super membrane (cut to the same size as the gel) in continuous buffer for 1 min and place on top of the filter paper 'trans-unit' (the stack of filter paper, gel, and membrane).

6 Allow the polyacrylamide gel to soak in continuous buffer for 5 min, then place it on top of the trans-unit and roll again to remove bubbles.

7 Wet another six pieces of filter paper by capillary action in continuous buffer solution. Place these on to of the trans-unit and once again roll to remove bubbles.

8 Calculate the total surface area to be electroblotted (more than one trans-unit can be processed simultaneously, but the gel/trans-unit areas must be added together).

9 Place the cathode (black/–) on top of the trans-unit. Pre-wet the graphite with distilled water as in step 4.

10 Allow the membrane to transfer for 60 min at 0.8 mA cm^{-2} of membrane. As some proteins transfer more or less rapidly than others two membranes can be used to capture protein 'blow through'.

Protocol 4

Detection by enhanced chemiluminescence

Equipment and reagents

- Blocking buffer: 1 g of non-fat dried milk powder 100 ml^{-1} phosphate-buffered saline (PBS)
- Nitrocellulose membrane
- Prediluted 1/1000 mouse monoclonal antibody
- 0.3% (v/v) Tween 20
- Anti-class antibody linked with horseradish peroxidase prediluted at 1/1000
- ECL detection reagent (Amersham Pharmacia Biotech)

Method

1 Following electrotransfer, remove the blot and label with a pencil in the top right-hand corner of the side supporting the transferred protein.

2 Place membrane in blocking buffer (1 g of non-fat dried milk powder 100 ml^{-1} PBS) at 4 °C overnight or for longer, if need be.

3 Place nitrocellulose membrane in a clean tray and wash × 5 in a shaker for 5 min each time in PBS containing 0.3% (v/v) Tween 20.

4 Add 20 ml of prediluted 1/1000 mouse monoclonal antibody and incubate with agitation at 37 °C for 1 h.

Protocol 4 continued

5 Repeat wash step 3.

6 Incubate with second antibody (anti-class linked with horseradish peroxidase pre-diluted at 1/1000) for 1 h with agitation at 37 °C.

7 Repeat wash step 3.

8 Cover the blot with 0.125 ml/cm^{-2} ECL detection reagent (Amersham Pharmacia Biotech). Leave for 1 min.

9 Remove blot, drain, and place a cellulose acetate folder/sac and heat sealed to avoid leakage. The pencil label (protein side is placed against a sheet of Hyperfilm-ECL (Amersham Pharmacia Biotech) within an X-ray exposure cassette.

4 Edman protein microsequencing

Protein microsequencing remains a reliable means of acquiring primary sequence information. This robust chemistry was first developed by Pehr Edman in 1949 (54); adapted to provide the basis of modern automated sequencers through the work of Edman and Begg in 1967 (55); modified to include covalent attachment of the C-terminal to a solid substrate by Laursen in 1971 (56); and finally rendered significantly more sensitive by the introduction of gas phase sequencing (57). Generally, the carboxy-terminal of a protein or peptide is rendered inaccessible to the Edman chemistries by binding to a solid substrate, then the free amino-terminal is processed one residue at a time and the protein sequence is called with respect to peak elution times from microbore reverse phase HPLC. This chemistry is maintaining its utility even in the age of advanced mass spectrometry of macromolecules, whereby it has found a competitive niche of relevance to the genomic sciences when employed to confirm start codons of small genes/open reading frames encountered during DNA sequencing initiatives (58). The first few cycles of Edman degradation chemistry also aid in sample clean-up and thus sufficient sequence information can often be obtained from contaminated samples in subsequent cycles so as to permit protein characterization. Of recent times, sensitivity has increased dramatically whereby 10–15 residues of sequence have been obtained from as little as 100 femtomole initial peak yield of protein (50, 58). This is similar to that reported for electrospray ionization mass spectrometry (ESI MS)/MS of proteins derived from 2-D electrophoresis gel (3); however, the latter has the advantage of superior sequence coverage due to information being derived from multiple peptides. The quantity of amino acid sequence obtained is strongly influenced by the yield from the first cycle of Edman chemistry and the repetitive yield thereafter. Recent improvements in this sector have allowed sequence calls to go beyond 70 amino acid residues (Simpson, pers. comm.).

The chemistry can be summarized as follows: (i) *coupling* of phenylisothiocyanate with the protein or peptide under basic conditions to form a phenylthiocarbamyl (PTC) peptide; (ii) *cleavage* of the end residue from the N-terminus of

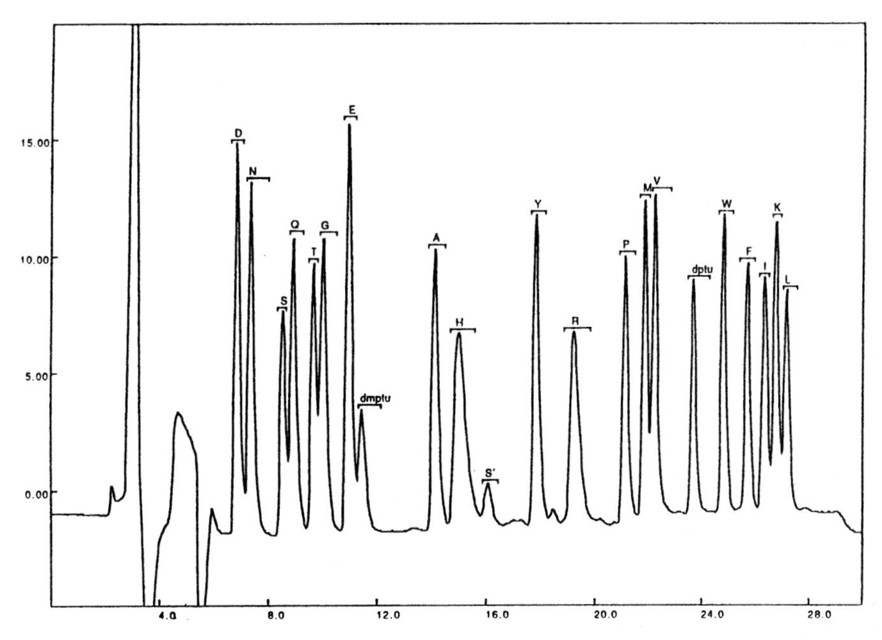

Figure 2 The elution profile of 19 PTH amino acid derivatives used as a standard for Edman protein microsequencing.

the protein or peptide using an anhydrous acid such as trifluoroacetic acid (TFA), so as to release an anilinothiazolinone (AZT) derivative of the amino acid residue in question; (iii) *conversion* of the AZT, again through the action of TFA, into a phenylthiohydantoin (PTH) amino acid derivative, which is eluted from a HPLC column. The retention time associated with the PTH derivative obtained from each repetition of the above cycle allows the operator to call the amino acid sequence of the peptide or protein (*Figure 2*).

Automated Edman sequencers can be obtained from a variety of commercial suppliers; however, good results still depend upon experienced operators optimizing machine performance and deciphering chromatograms, particularly as the concentration of the initial sample or PTH derivative decreases. Thus, learning the trade from an experienced user is the best recommendation one can give to beginners. The reader is referred to the following reviews for a more detailed analysis of protein microsequencing (59–61).

Sample impurities (such as, multiple polypeptides, salts, detergents, and free amino acids) can confuse the results obtained. However, a major complication with this procedure is N-terminal blockage from both biological and artificial sources. The agents responsible for blocking have been reviewed by us previously (9), while the simplest mechanism to overcome this problem is the cleavage of the protein to create new N- and C-termini corresponding to internal fragments (see *Protocol 9*). Chemical cleavage with cyanogen bromide (see *Protocol 10*) and a variety of endo- and exo-peptidases can be used for the generation of internal fragments (62, 63, see list of enzymes and protocols later). Nonetheless, the following serve to reduce the risk of artificial blockage at the N-terminus due principally to acylation and methylation: (i) ageing the SDS–polyacrylamide gels prior to use; (ii) addition of thioglycolic acid or sodium thiosulphate to running buffer and protein sample to scavenge free radicals and oxidants (64); (iii) use of 3-[cyclohexylamino]-1-propanesulphonic acid (CAPS) rather than Tris/glycine buffers during electrotransfer (65); (iv) minimizing exposure to acetic acid during spot detection on polyvinylidene difluoride (PVDF) membranes (64); (v) use of Amido Black in preference to Coomassie Brilliant Blue R250 for PVDF membrane staining (66, 67); and (vi) drying membranes under nitrogen to reduce the likelihood of oxidation.

When interrogating protein databases against a peptide string, it is very important to conduct searches using, for example, *FastA* and/or *trans FastA* against: (i) the entire database, and (ii) a modified database containing only N-termini. The former must be included so as to detect the sequence corresponding to stable cleavage products and the latter so as to screen a much larger pyramid of phylogenetically similar entries independent of the noise contributed by internal sequence similarity. The search directed against only N-terminal sequence is critical for proteins derived from organisms not yet fully sequenced at the level of DNA.

Protocol 5

Preparation for protein microsequencing (and amino acid analysis or mass spectrometry)

Equipment and reagents

- 2-D electrophoresis gel
- PVDF membrane (Problot, Applied Biosystems)
- Transfer buffer from *Protocol 3* replaced with: 10% CAPS 10 × Stock, 10% methanol
- 2 M NaOH
- Problott™ membrane
- Amido Black or Ponceau-S
- 30% (v/v) acetonitrile

Method

1. Transfer proteins resolved from within a well-focused 2-D electrophoresis gel to a PVDF membrane (Problot, Applied Biosystems) using *Protocol 3*, except replacing the transfer buffer with the following (10% CAPS 10 × Stock and 10% methanol in water; CAPS stock solutions = 100 mM solution adjusted to pH 11.0 using 2 M NaOH) and extending the transfer time to 120 min. We have found that Problott™ membrane produces a desirable reduction of protein 'blow through'.
2. Remove the PVDF membrane from the 'trans-unit' and stain the membrane according to either *Protocol 6* or *Protocol 7*. (Revelation of proteins using Amido Black or Ponceau-S gives fewer background peaks and reduces the problem associated with Coomassie Blue staining of blots, whereby proteins can be lost during destaining.)
3. Excise protein spot with a clean scalpel blade and place in an microcentrifuge tube.

Protocol 5 continued

4 Add 1 ml of 30% (v/v) acetonitrile in water and vortex for 1 min.
5 Discard acetonitrile and repeated the vortex wash for 1 min.
6 Discard acetonitrile and dry under nitrogen filtered at 0.2 μm and streaming gently from a Pasteur pipette into the microcentrifuge tube.
7 When dry, seal the tube and place it in a sac filled with filtered nitrogen. The specimen can now be held or dispatched until ready for microsequencing.

Note: The signal-to-noise ratio here is more critical than the quantity of protein available for sequencing. Thus a small amount of protein focused into a small surface area can yield excellent results. Provided endo-electro-osmosis is avoided, focusing times in the first dimension of gel electrophoresis should be extended to at least 60 000–100 000 V/H.

Protocol 6

Amido Black staining of PVDF membrane

Equipment and reagents

- PVDF membrane
- Naphthaline Black 10B 0.1% (w/v)
- Isopropanol 25% (v/v)
- Glacial acetic acid 10%

Method

1 Following electrotransfer of proteins, wash PVDF membrane twice with distilled water for 5–10 min accompanied by shaking.
2 Stain membrane with solution containing Naphthaline Black 10B 0.1% (w/v); isopropanol 25% (v/v); and glacial acetic acid 10% 5 min.
3 Destain in distilled water for approximately 5–10 min.

Protocol 7

Ponceau-S staining of PVDF membrane

Equipment and reagents

- 0.1% Ponceau-S
- 5% (v/v) acetic acid
- Distilled water
- 200 μM NaOH in 20% (v/v) acetonitrile

Method

1 Stain with 0.1% Ponceau-S in 1% (v/v) acetic acid for 5 min.
2 Destain in 5% (v/v) acetic acid for 5–10 min.
3 Wash twice with distilled water for 5 min accompanied by shaking.

Protocol 7 continued

4 Destain with 200 μM NaOH in 20% (v/v) acetonitrile for 5–10 min.

5 Wash twice with distilled water for 5 min accompanied by shaking.

5 Amino acid composition

Amino acid analysis is based upon the non-ordered composition of a protein and as such is less sensitive to amino acid substitution events across species boundaries or minor experimental inaccuracies. Indeed, this measure is highly desirable during protein characterization, especially when combined with other orthogonal analyses, such as, peptide mass fingerprinting (PMF) and/or 'sequence tagging' (68). Unfortunately, the process requires at least 5 pmol of sample, i.e. only a minor portion of proteins displayed on a 2-D electrophoresis gel. In addition, the accuracy of the measure is severely compromised by inadequate acid hydrolysis of some residues and partial degradation of others; and the error introduced by amino acids found in the general environment and employed in buffer solutions. The experimental procedure has been reviewed by several authors (9, 69, 70). It maintains currency due to the fact that, even across species boundaries, a protein rich in proline, for example, will remain so through evolutionary time and independent of a certain degree of experimental inaccuracy.

Protocol 8

Analysis of amino acid composition

Equipment and reagents

- PVDF membrane
- HPLC vial
- Hydrolysis vessel ('Sputnik')
- Hydrolysis solution: 6 N HCl, 1.0 % phenol
- Elution solution: 60% acetonitrile, 0.1% trifluoroacetic acid
- 2.5 mM 4-hydroxy-L-proline in 0.1 N HCl

Method

1 Excise protein spot electroblotted on to PVDF membrane and cut this into pieces approximately 1 × 1 mm.

2 Place these membrane pieces into a small glass 'micro-insert', that fits within a standard HPLC vial.

3 These micro-inserts are then placed within a Teflon block containing up to 100 wells compatible with the external diameter of these micro-inserts.

4 Place one or more Teflon blocks within the hydrolysis vessel ('Sputnik'; see *Figure 3*) containing 10 ml of hydrolysis solution (6 N HCl; 1.0 % phenol).

Protocol 8 continued

5. Close the hydrolysis vessel taking particular attention to verify that the heat tolerant ring-seal is properly positioned. A torsion wrench is then employed to ensure similar pressure is applied across the six bolts sealing the two glass halves of the 'Sputnik'.
6. Connect the hydrolysis vessel to an inert gas supply (Argon) and a Teflon-lined vacuum pump via a three-way tap.
7. Eliminate all oxygen from the hydrolysis vessel by repeated flushing with the inert gas and subsequently evacuating the hydrolysis vessel using the vacuum pump. Note that when flushing the hydrolysis vessel, one should not allow the pressure to exceed 200 kPa.
8. Repeat step 7 at least six times. On the final occasion, allow evacuation to continue until the resultant vacuum causes the acid hydrolysis solution to bubble.
9. Close the three-way tap to maintain vacuum prior to turning the vacuum pump off.
10. Place hydrolysis vessel in a convection oven for 24 h at 110 °C.
11. Transfer hydrolysis vessel to fume hood, release acid vapour and allow to cool.
12. Remove the Teflon block containing glass micro-inserts.
13. Dispense 20 μl of elution solution (60% acetonitrile, 0.1% trifluoroacetic acid) or sufficient volume to immerse membrane pieces.
14. Place each micro-insert into a 2.5 ml microcentrifuge tube and spin the tubes to ensure that PVDF pieces are completely submerged.
15. Incubate overnight at 37 °C.
16. Sonicate for 15 min three times in a water bath at room temperature.
17. Dry down samples by placing the microcentrifuge tubes containing micro-inserts in a rotary evaporator.
18. Add 1 μl of 2.5 mM 4-hydroxy-L-proline in 0.1 N HCl, as an internal standard for HPLC.
19. Replace micro-inserts into standard 2 ml HPLC vials and seal vial with a lid containing a piercable septum.
20. The samples are now ready for chromatographic separation. Elution times being compared with a BSA standard.

Note: A GBC Aminomate HPLC system with autosampler; C18 reverse phase column and pre-column FMOC derivatization is used in conjunction with an on-line fluorescent detector. The mobile phase consists of three solutions; 2 M phosphate buffer (pH 6.5); 15% methanol; and 90% acetonitrile. Ongoing efforts in our laboratory are directed at selecting the barest minimum of experimentally reliable residues to achieve statistically accurate characterization of proteins during database interrogations. The need for this improvement is necessary, as far too many amino acid residues vary excessively with respect to hydrolysis run, amino acid composition of the sample, molecular mass and run-to-run variation during HPLC. (Method supplied by D. Basseal, pers. comm.)

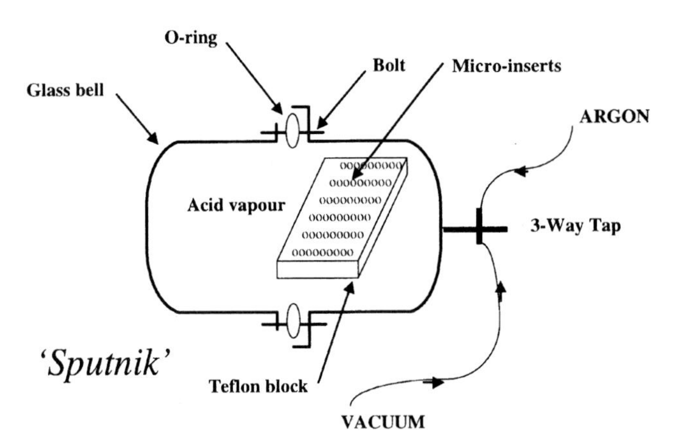

Figure 3 Acid hydrolysis vessel employed for large-scale processing of 1–200 protein samples.

6 Proteinase cleavage, sample concentration, clean-up, and preparation for mass spectrometry

Initially, proteins resolved within polyacrylamide gels must be stained, so that they can be excised and further characterized. A variety of approaches have been reviewed (71).

Whether to remove blocked N- or C-termini from proteins for Edman microsequencing (*Protocol 9*) or for the generation of peptides for more detailed analysis by mass spectrometry, endoproteinase digestion/cleavage (*Table 1*) or chemical fragmentation (*Protocol 10*) are often among the first laboratory manipulations conducted on purified proteins. The significance of these approaches during protein characterization by PMF will be discussed later. Of relevance to protein digestion, and indeed protein labelling, is the incidence of a particular amino acid residue targeted for cleavage (*Table 2*).

Table 1 Commonly employed endoproteinases

Enzyme	Cut site	Digestion buffer[a]	Enzyme to substrate	Digestion at 37°C	Resists	pH range
Trypsin	R–X K–X	100 mM ammonium bicarbonate	1 : 50–100	2–16 h	0.1% SDS	7.0–90
Glu-C(i)	E–X	100 mM ammonium bicarbonate	1 : 30	2–18 h	4 M Urea 0.2% SDS	4.0–8.0
Glu-C(ii)	E–X D–X	100 mM phosphate	1 : 30	2–18	4 M Urea 0.2% SDS	7.8
Lys-C	K–X	100 mM ammonium bicarbonate	1 : 50	22 h	5 M Urea 0.5% SDS	8.5–8.8
Asp-N	X–D	50 mM phosphate	1 : 20	4–18 h	2 M Urea	7.0
Arg-C	X–R	100 mM ammonium bicarbonate	1 : 50	8 h	0.1% SDS 3 M Urea	7.5–8.5

[a]In all cases approximately 10 g l^{-1} of substrate is dissolved into the digestion buffer.

Table 2 Properties of amino acid residues

Residue	3-letter code	1-letter code	Residue mass	Mass of most abundant imonium ion	Mass of related ions	Abundance of residue	Absence of residue
Alanine	Ala	A	71	44	–	7.40[a]	2.69[b]
Arginine	Arg	R	156	129	59, 70, 73, 87, 100, 112	5.20	5.15
Asparagine	Asn	N	114	87	70	4.59	6.27
Aspartic Acid	Asp	D	115	88	70	5.18	5.38
Cysteine	Cys	C	103	76	–	1.8	16.72
Glutamic Acid	Glu	E	129	102	–	6.21	5.09
Glutamine	Gln	Q	128	101	56, 84, 129	4.07	6.26
Glycine	Gly	G	57	30	–	7.00	2.74
Histidine	His	H	137	110	82, 121, 123, 138, 166	2.24	13.99
Isoleucine	Ile	I	113	86	44, 72	5.59	5.23
Leucine	Leu	L	113	86	44, 72	9.16	2.79
Lysine	Lys	K	128	101	70, 84, 112, 129	5.79	5.18
Methionine	Met	M	131	104	61	2.30	11.12
Phenylalanine	Phe	F	147	120	91	4.00	6.05
Proline	Pro	P	97	70	–	5.06	5.11
Serine	Ser	S	87	60	–	7.39	2.67
Threonine	Thr	T	101	74	–	5.97	3.45
Tryptophan	Trp	W	186	159	77, 117, 130, 132, 170, 171	1.34	20.37
Tyrosine	Tyr	Y	163	136	91, 107	3.26	7.57
Valine	Val	V	99	72	41, 55, 69	6.45	3.95

[a] Percentage of total known proteins as per Wise *et al.* (86). (NB These figures are not expected to vary significantly from the calculations made on the 128 719 proteins contained in the OWL non-redundant protein database as in June 1995.)

[b] Percentage of a non-redundant protein database entries OWL (86). This information is critical to the likely success of protein cleavage and the isotopic or fluorescent tagging of amino acid residues. Noteworthy are W, M, H and C, which are absent from a significant percentage of all known proteins.

However, all too often in protein chemistry, one is confronted with insufficient sample (protein or peptides) to conduct effective characterization. Under such circumstances, a number of options are available to increase the concentration of the protein sample and thereby the likelihood of obtaining adequate data from analytical methods. Although far from exhaustive, the following protocols should prove useful towards this end. The traditional approach of protein precipitation (*Protocol 11*) should not be forgotten here, yet it is only recommended for use with more abundant proteins contained in larger volumes. Similarly, electro-elution of proteins is also practicable on protein samples that have been passively eluted from acrylamide gels (*Protocol 12*). Protocols for electro-elution are best conducted following those recommended by the appropriate commercial manufacturer. More efficient still is the approach of spinning down passively eluted proteins on to a small piece of ProBlott™ PVDF membrane contained within a ProSpin™ column (Applied Biosystems) or combining several identical gel bands or spots so as to increase the abundance of the protein of interest (*Protocol 13*).

In addition, elution of individual peaks from a variety of HPLC columns provides a means of further concentrating the analyte within a reduced volume. Although HPLC methods will not be addressed here, there exists a wonderfully simplified approach of combining the affinity of a sieving matrix for protein binding during HPLC with protein concentration, clean-up and matrix-assisted laser desorption/ionization time-of-flight mass spectrometry (MALDI-TOF MS) (*Protocol 14*).

For extremely low abundance proteins, affinity chromatography on-line during fractionation by capillary zone electrophoresis and delivery to mass spectrometry (or other analytical platform) is increasingly seen to hold much promise. Here, peptide generation can also be conducted on-line prior to peptide concentration using, for example, antibodies directed against peptides labelled with a suitable chromophore. The reader is referred to several recent reviews in the field of analyte concentration, solid phase extraction, and affinity capillary electrophoresis (72–75).

Protocol 9

Peptide microsequencing following in-gel digestion

Equipment and reagents

- Stained polyacrylamide gel
- 0.2 M ammonium bicarbonate (AMBIC)/50% acetonitrile
- 0.2 M AMBIC/0.02% Tween-20
- 10% TFA
- Narrow-bore reverse phase HPLC

Method

1. Commence with a well-reduced and alkylated sample excised from a stained polyacrylamide gel. (See note *Protocol 17* regarding importance of sample alkylation and reduction.)

Protocol 9 continued

2 Wash twice for 20 min at 30 °C with 100–200 μl of 0.2 M AMBIC/50% acetonitrile.
3 Dry *completely* under a slow stream of nitrogen.
4 Rehydrate with 2–5 μl of 0.2 M AMBIC/0.02% Tween-20.
5 Add protease (modified trypsin works well) in 5 μl.
6 Continue rehydration with small aliquots of 0.2 M AMBIC/0.02% Tween-20 until the gel piece is fully reswollen.
7 Incubate overnight at 30 °C.
8 Acidify with 1/10 volume of 10% TFA.
9 Extract the peptides generated for at least 30 min twice at 30 °C with about 150 μl of 0.1% TFA/60% acetonitrile and combine supernatants.
10 Evaporate the organic phase in a rotary evaporator.
11 Isolate peptides by narrow-bore reverse phase HPLC, prior to microsequencing on glass, nitrocellulose, PVDF, or on-line to mass spectrometry.

Note: Use proteins stained by protocols not employing more than 5% methanol in the fixing step (76).
Method supplied by U. Hellman and after methods (76, 77).

Protocol 10

Chemical fragmentation with cyanogen bromide

Equipment and reagents

- PVDF or nitrocellulose membrane
- 70% formic acid
- Cyanogen bromide solution
- Distilled water
- 0.1% TFA
- microbore HPLC

Method

1 Cut protein spot or band which has been electrotransferred on to PVDF or nitrocellulose into small piece approximately 1 mm × 1 mm and place these in a microcentrifuge tube.
2 Suspend these in 100–200 μl of 70% formic acid (sufficient to cover all sample pieces).
3 Add approximately 10 μl per 10 μg of protein of cyanogen bromide solution (70 mg ml^{-1} of 70% formic acid) or a supposed excess, i.e. 10 μl of above will suffice for most 2-D electrophoresis samples.
4 Incubate in the dark at room temperature for 24 h.
5 Remove the cleavage solution.

Protocol 10 continued

6 Dilute with approximately 10 volumes of distilled water.
7 Dry in a rotary evaporator.
8 Redissolve in 0.1% TFA and prepare for microbore HPLC.

Protocol 11

TCA precipitation

Equipment and reagents

- Protein sample
- 60% (v/v) trichloroacetic acid
- Acetone

Method

1 Place the solution containing protein sample and a 60% (v/v) trichloroacetic acid (TCA) solution on ice for 20 min.
2 Add four volumes of protein sample to one volume TCA solution.
3 Leave on ice for 120 min.
4 Centrifuge at 4000 g or more in a refrigerated centrifuge for 15 min.
5 Remove supernatant leaving some liquid in the bottom of centrifuge tube.
6 Wash three times using 100 μl of acetone at 4 °C. Note that the precipitate may not be visible following steps 5 and 6.
7 Resuspend in a reduced volume of an appropriate buffer for subsequent analysis.

Protocol 12

Passive elution of proteins from polyacrylamide gels

Equipment and reagents

- Polyacrylamide gel
- 0.2% Coomassie brilliant blue R250 (w/v)
- 20% isopropanol (v/v)
- 0.5% glacial acetic acid (v/v)
- 79.5% deionized water
- 30% methanol (v/v)
- Elution buffer: 100 mM sodium acetate (pH 8.5), 0.1% SDS (w/v)

Method

6 Incubate at 37 °C for 10–16 h.
7 Take-up protein or peptides contained within elution buffer.
1 Stain the polyacrylamide gel for as short a time as possible in 0.2% Coomassie brilliant blue R250 (w/v), 20% isopropanol (v/v), 0.5% glacial acetic acid (v/v), and 79.5% deionized water, i.e. until the gel takes on a uniform blue appearance.

Protocol 12 continued

2 Destain in 30% methanol (v/v) in deionized water.
3 Excise the spots or bands of interest with a sterile scalpel blade and place these in a microcentrifuge tube.
4 Wash in deionized water for 30 min. Remove water.
5 Add 300 μl of elution buffer: 100 mM sodium acetate (pH 8.5) and 0.1% SDS (w/v) in deionized water.
6 Incubate at 37 °C for 10–16 h.
7 Take-up protein or peptides contained within elution buffer.

Note: Best results are achieved in the absence of higher concentrations of glacial acetic acid during staining and destaining. It is desirable to reduce dramatically the volume of elution buffer so as to merely cover the gel pieces with sufficient liquid to leave a small volume, for example, 1–5 μl at the end of step 8.

Protocol 13

Digestion, concentration, and desalting of protein contained in multiple gel spots or bands

Equipment and reagents

- 200 mM Tris-HCl (pH 8.5)
- 10% acetonitrile
- 1 mM $CaCl_2$ (pH 8.5)
- Trypsin
- 1% and 2% TFA (v/v)
- MALDI-TOF MS
- HPLC
- ESI MS

Method

1 Gel pieces corresponding to several protein spots or bands are cut in to small pieces approximately 1 mm × 1 mm.
2 Gel pieces are shrunk in a sufficient volume of a mixture 1:1 of 200 mM Tris-HCl at pH 8.5 and acetonitrile, so as to cover all the gel pieces. This is held at 30 °C for 30 min with shaking.
3 Wash solution is discarded and gel pieces are rehydrated in a minimal volume of 100 mM Tris-HCl buffer, i.e. just enough to moisten all the gel pieces.
4 Gel pieces and Tris-HCl buffer are then transferred to a 1-ml syringe and passed through a fine stainless steel sieve (mesh size approximately 100 μm) to produce a gel slurry extruded in to a microcentrifuge tube.
5 Evaporate liquid in a rotary evaporator, without drying completely.
6 Reswell gel slurry in 50–100 μl of 100 mM Tris-HCl; 1 mM $CaCl_2$ at pH 8.5 and 0.2–0.5 μg of trypsin in 10% (v/v) acetonitrile for 30 min to initiate digestion.

Protocol 13 continued

7. Add a further 1–5 volumes of buffer, as is step 6, and incubate for 15 h at 36 °C.
8. Add two volumes, with respect to the total present in step 7 of 2% TFA (v/v) in water and hold at 60 °C with shaking for 1 h.
9. During step 8, prepare a reverse phase column in a disposable pipette tip containing 3–5 mg of reverse phase material or employ Zip-Tips™, as in *Protocol 14*. (Pipette-tip column is washed with 1 ml of 60% acetonitrile/0.1 % TFA and equilibrated with 1 ml of 0.1% TFA ready for step 11.)
10. The gel slurry from step 8 is poured into a 2 ml syringe forming part of the peptide-collecting device (PCD), as shown in *Figure 4*.
11. The liquid phase derived from the gel slurry is now slowly pressed though the PCD and the eluate discarded.
12. The pipette-tip column is removed and the gel slurry combined with 1–2 volumes of 0.1% TFA.
13. Repeat step 11 and remove pipette-tip column.
14. Remove the glass wool packing and the gel pieces from the pipette-tip column.
15. Fix the pipette-tip column to a new syringe and wash with 1–2 ml of 0.1% TFA (discarding eluate).
16. Elute peptides into a microcentrifuge tube using 300 μl of 0.1% TFA in 60% acetonitrile.
17. 5% of sample is directed to MALDI-TOF MS and the remained placed in a rotary evaporator and the volume reduced to approximately 50–60 μl before HPLC in conjunction with protein microsequencing or ESI mass spectrometry (see later).

Method after Otto *et al.* (78).
Note: Similarly efficient methods for concentrating protein from multiple gel spots or band have also been described elsewhere (79–81).

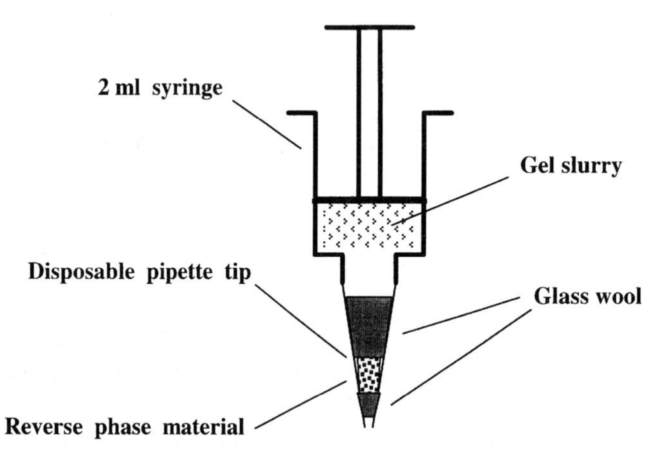

Figure 4 Syringe funnel used to concentrate protein derived from several polyacrylamide gel bands or spots, after ref. (78).

Protocol 14

Microscale concentration and desalting of peptide mixtures prior to mass spectrometry

Equipment and reagents

- ZipTip™ pipette tip (Amicon/C18 silica, 15 μm, 200 Å pore size, Millipore, Product No. ZTC 18S 096)
- Resin
- 50% acetonitrile
- 0.1% TFA
- 5% methanol

Method

1. Use a 10 μl disposable ZipTip™ pipette tip (C18 silica, 15 μm, 200 Å pore size) with approximately 0.6 μl bed of resin fixed at its end intended for concentrating, desalting, and removing detergents from biological samples.
2. Pre-wet the tip by depressing plunger to a dead stop using the maximum volume setting on a 10 μl micropipette. Aspirate wetting buffer (50% acetonitrile in water) into tip. Dispense to waste.
3. Equilibrate the tip for binding by washing it twice with the equilibration buffer (0.1% TFA in water).
4. Bind peptides or small proteins (<40 kDa) to ZipTipC18 by fully depressing the pipette plunger to a dead stop. Aspirate and dispense peptide sample five to 10 times depending on the sample concentration. (Dilute solutions require increased contact time.)
5. Wash tip and dispense to waste using two cycles of 0.1% TFA in water; or 5% methanol in 0.1% TFA/water mixture. (Electrospray applications or samples containing high salt or detergent may require additional washing.)
6. Elute peptides by dispensing 2–4 μl of elution buffer (50% acetonitrile in water) into a clean microcentrifuge tube using a standard pipette tip. Carefully, aspirate and dispense eluant through ZipTip at least three times without pushing air into the sample. For ESI MS, elute sample into clean vial and apply directly for analysis. For MALDI MS, elute with or without matrix in 50% acetonitrile/water by pipetting 1–2 μl of desalted-concentrated sample directly on to MALDI-TOF target by depressing plunger until appropriate volume is dispensed (see also *Protocol 19*).
7. Save or discard the remaining sample with tip, i.e. sample can be eluted at a later date from the sample tip.

Notes: As the adsorptive bed provides back pressure, do not use ZipTip for accurate volumetric dispensing. For direct MALDI, prepare matrix in 50% acetonitrile/water and elute peptides directly on to target. For samples containing 1 μg of protein or 0.5 μg of peptide, you can achieve maximum recovery by increasing elution volume to 10 μl or by performing multiple elutions with 5 μl of elution buffer. For pipetting steps, press the

Protocol 14 continued

ZipTip securely on to pipette and depress plunger to dead stop to allow maximum suction and sample dispensing.
Maximum binding to the ZipTip is achieved in the presence of TFA or other ion-pairing agents. To minimize sample dilution and enhance analyte binding, use 10% TFA to achieve a final concentration of 0.1% TFA. In the case of excess detergent, dilute sample with 0.1% TFA to achieve acceptable binding conditions, for example, SDS (<0.1%), Triton® (<1%), and Tween® (<0.5%).

Protocol 15

Bead concentration for delivery to MALDI-TOF MS

Equipment and reagents

- Protein sample/reverse-phase HPLC eluate
- 0.1% (v/v) TFA
- Suspension of Poros R2 beads in absolute methanol
- MALDI matrix solution: 40 mg α-cyano-4-hydroxycinnamic acid, 10 ml 2,5-dihydrobenzoic acid in 500 μl 0.1% (v/v) TFA in water/acetonitrile

Method

1. Take protein sample or reverse-phase HPLC eluate and dilute in at least five volumes in 0.1% (v/v) TFA in water containing 5 μl of a 1 mg ml^{-1} suspension of Poros R2 beads in absolute methanol.
2. Incubate for 10 min in a microcentrifuge tube with occasional mixing.
3. Centrifuge for 10 min in a benchtop microcentrifuge at 10 000 *g*.
4. Discard all supernatant leaving behind 3 μl in bottom of tube.
5. Transfer pelleted beads (to which peptides have now bound) to a clean MALDI-TOF target and let air dry.
6. Elute with 0.5–0.7 μl of MALDI matrix solution (a fresh fivefold dilution of 40 mg α-cyano-4-hydroxycinnamic acid and 10 ml 2,5-dihydrobenzoic acid in 500 μl 0.1% (v/v) TFA in water/acetonitrile) and let air dry.
7. Perform MALDI-TOF MS taking care to avoid firing the laser directly at the Poros beads.

Method supplied by K. Gevaert, H. De Mol and J. Vandekerckhove (pers. comm.).

7 Mass spectrometry

7.1 Hierarchical analysis

Over the past decade, mass spectrometry has revolutionized the means by which proteins are characterized (82). Protein characterization by mass spectrometry

Table 3 A schematic representation of the '*combined approach*' employed using two endoproteinases during protein characterization

Ranked output	Trypsin	Glu-C
1 (most similar)	Protein AAAA	Protein PPPP[a]
2	Protein BBBB	Protein AAAA
3	Protein CCCC	Protein RRRR
4	Protein DDDD	Protein SSSS
5	Protein EEEE	Protein TTTT
6	Protein FFFF	Protein UUUU
7	Protein GGGG	Protein VVVV
8	Protein HHHH	Protein JJJJ
9	Protein IIII	Protein WWWW
10 (least similar)	Protein JJJJ	Protein XXXX

Combined Score for Protein AAAA is $1 \times 2 = 2$.

Combined Score for Protein JJJJ is $10 \times 8 = 80$.[b]

Lowest score is associated with the greatest statistical confidence.

Without additional attributes, other entries must be considered as non-reliable i.e. lack of co-occurrence.

[a] Further columns can be added to such tables that correspond to still further analytical procedures, such as, additional endoproteinase digestion, amino acid composition and N-terminal sequencing, whereby co-occurrence in lists derived from database interrogation increases the statistical confidence associated with a protein characterization.

[b] Scores of <100 are highly significant for two orthogonal analytical procedures. Further confidence in the results obtained can be preened by comparison with the expected M_r and pI as determined from the *x/y* co-ordinates on 2-D electrophoresis gels and the likelihood of enzymatic presence in metabolic pathway predictions for the a given organism.

follows an hierarchical approach based upon decreasing speed of analysis and concomitantly an increasing degree of operator interpretation required for spectral analysis, although here too dramatic improvements are being realized. In all cases, this hierarchy is dependent upon peptide mass fingerprinting (PMF) combined with respectively: (i) orthogonal datasets such as additional PMF produced by alternate enzymes (see *Table 3*), amino acid composition and/or total molecular mass; (ii) 'sequence tagging' (*Figure 5*) (83) or predicted fragmentation patterns using SEQUEST software (*Figure 6*) (84); and (iii) *de novo* sequencing by ESI-MS, so as to provide sufficient information for the direct targeting of genes of interest by nucleotide probes, degenerate PCR and/or 'in silico' cloning with respect to expressed sequence tags (EST) databases. The above methodologies are based upon replacing the chemical code of proteins with 'unique numerical parameters', which can thereby facilitate unambiguous annotation of gene products. As the number of database entries corresponding to fully sequenced (DNA) organisms and ESTs increases, then so too does the likelihood of rapid protein identification. Yet, *de novo* sequencing will remain in demand for some time to come.

7.2 Peptide mass fingerprinting

In most cases, the initial step in protein characterization by mass spectrometry involves the generation of a peptide fingerprint with respect to the molecular

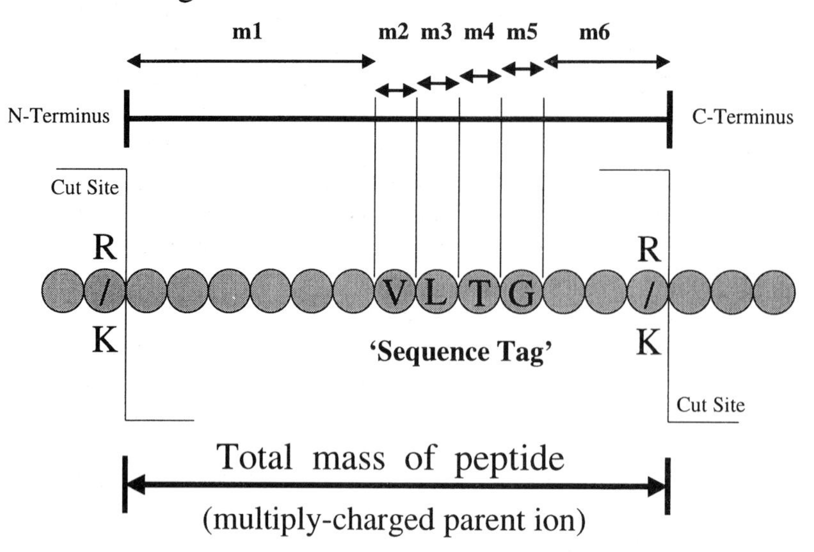

Figure 5 'Sequence tagging' of a peptide, after ref. (83).

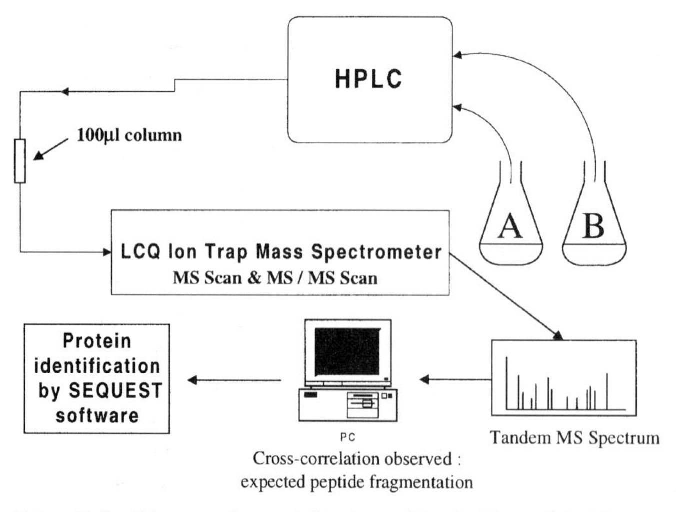

Figure 6 Peptide mass fingerprinting is combined with predicted fragmentation patterns and SEQUEST software during protein characterization, after ref. (84).

mass of the peptides produced following enzymatic digestion or chemical fragmentation. As such PMF is essential to all aspects of proteome analysis. This led Wise *et al.* (85) to examine the relative limitations and efficiencies of the procedure based upon the analysis of some 51 million peptides generated by perfect 'in silico' cleavage using 23 different enzymes. Characterization is dependent upon proteins being able to be cleaved at particular sites (for example, the absence of a particular residue from a protein of interest) and the incidence of these cut sites with respect to the production of statistically significant peptides of value in protein characterization (*Table 2*). This field has been reviewed recently (29). Experimental data will always be inferior to such perfect data, the latter being compromised by inefficient cleavage due to poor alkylation and reduction, the M_r of the protein, sample impurities, the desorption matrix, chemical adducts, sample concentration, column eluate, or digestion buffer. The 'Combined Approach' (86) can provide statistically valid characterization for the bulk of known proteins (*Table 3*). Here we recommend the use of two endoproteinases, such as, trypsin and Glu-C. Detection of a similar PMF then allows the attribution of 'gene-product family' status to several proteins encoded by the same gene, i.e. possessing similar 'unique numerical attributes' and of critical importance to dissecting gene and genomic function *in vivo*. All potential peptides generated from a protein are rarely detected, and thus protein characterization by a combination of PMFs is usually dependent upon the detection of only a few non-modified peptides that are structurally or functionally significant to the protein or protein family (87).

7.3 MALDI-TOF MS

In 1988 two groups independently achieved 'soft ionization' of protein macromolecules. This included the development of electrospray ionization (88–90) and MALDI-TOF (91) mass spectrometry. The latter offers the greatest hope for eventually achieving high throughput proteomics on a genomic scale, while in just over one decade it has evolved into a very user-friendly platform for biologists with little or no previous background in mass spectrometry. Protein samples are introduced into the mass spectrometer within or in juxtaposition with a crystalline matrix and are then ionized most often through the action of a nitrogen laser (337 nm), but infra-red laser can also be employed (92). The matrix absorbs UV light causing it to vaporize and ionize and in so doing causes the polypeptide to be introduced into a flight tube. The masses of polypeptides are then measured with respect to the time taken for ions to reach the detector since the initiating event, i.e. the precisely timed laser pulse (*Figure 7*). Recent improvements have included delayed extraction, so as to provide greater unison in the 'take-off' of a set of ions. The time-of-flight allows mass-to-charge to be calculated with unsurpassed resolution dependent upon the length of the flight path, while sensitivity is afforded by the efficiency of ion transmission. Internal standards can be added or trypsin autolysis products employed during PMF for mass calibration. The use of a reflectron device allows for post-source decay spectra to be generated and concomitantly the hope of high throughput sequence tagging of proteins on a scale more relevant to genomic screening. MALDI-TOF MS also offers great utility for the determination of total molecular mass of intact proteins. This information, when combined with PMF data, provides a reliable tool for the characterization of non-modified gene-products

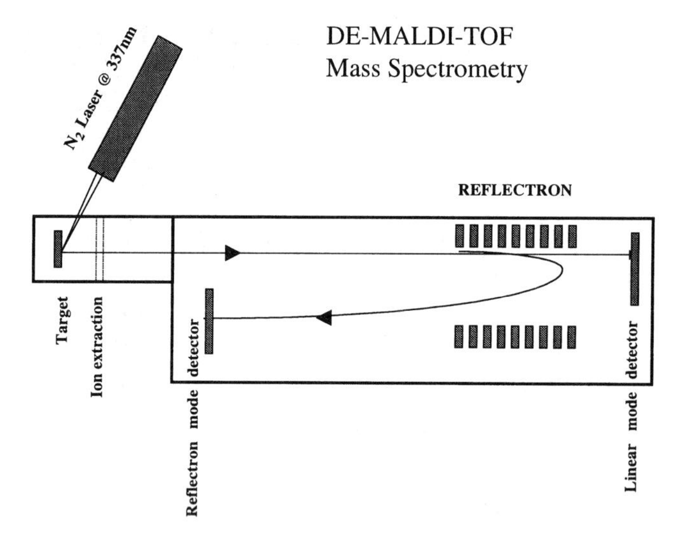

Figure 7 A schematic representation of MALDI-TOF MS.

in fully sequenced (DNA) organisms. *Protocols 16–19* are sufficient to obtain such data.

Protocol 16

Silver staining prior to analysis of proteins by mass spectrometry

Equipment and reagents

- PAGE gel
- 50% (v/v) methanol
- 5% glacial acetic acid (v/v)
- Distilled water
- 0.02% sodium thiosulphate
- Silver nitrate 0.1% (w/v)
- 0.04 % (v/v) formalin
- 2% (w/v) sodium carbonate solution

Method

1. Fix PAGE gel in 50% (v/v) methanol; 5% glacial acetic acid (v/v) for 1 h at room temperature.
2. Wash in 50% (v/v) methanol overnight (or for 1 h, if no sodium thiosulphate in electrophoresis gel).
3. Wash in three times in distilled water for 10 min at room temperature.
4. Desensitize in 0.02% sodium thiosulphate twice for 15 min (or 16–24 h, if no sodium thiosulphate in electrophoresis gels).
5. Wash three times in distilled water for 10 min at room temperature.
6. Submerge in chilled (4 °C on ice for at least 1 h in advance) silver nitrate 0.1% (w/v)

Protocol 16 continued

solution and incubate with shaking at 4°C for 1–2 h (able to be increased for convenience).

7. Rinse twice one-gel-at-a-time in chilled (4 °C on ice for at least 1 h in advance) distilled water for exactly 1 min.
8. Develop one gel at a time in 0.04 % (v/v) formalin (35% (v/v) formaldehyde in water is equivalent to concentrated formalin) in 2% (w/v) sodium carbonate solution at room temperature.
9. Discard developer after it turns yellow and replace with fresh developer.
10. Discard developer and stop with 5% glacial acid for 15 min.
11. Wash in distilled water for 30 min at room temperature and store at 4 °C in a sealed sac with approximately 10 ml of distilled water.

Notes: If conducted correctly, this method will produce wonderfully clear background to silver-stained spots. Silver staining is not generally compatible with mass spectrometry and/or endoproteinase digestion; however, the above method relies upon staining only some of the protein exposed at the gel surface, thereby maintaining some of the sensitivity of protein detection by silver, while leaving some of the protein intact for subsequent endoproteinase digestion and analysis by mass spectrometry.
(After procedure described elsewhere (93) and modified by M. Smith, H. Wilson, and IHS, pers. comm.).

Protocol 17

Reduction/alkylation and tryptic digestion 'in-gel' of silver-stained protein spots

Equipment and reagents

- Polyacrylamide gels
- Distilled water
- Destain solution: 30 mM potassium ferricyanide, 100 mM sodium thiosulphate
- Dithiothreitol solution: 6.5 mM dithiothreitol in 50 mM ammonium bicarbonate (pH 8.5)
- Acetonitrile
- Iodoacetamide solution: 54.0 mM iodoacetamide in 50 mM ammonium bicarbonate (pH 8.5)
- Enzyme stock: 0.1% trifluoroacetic acid (TFA), 100 mg of solid enzyme, 50 mM ammonium bicarbonate (pH 8.5)

Method

1. Protein spots are excised from polyacrylamide gels and placed into individual wells in a 96-well microtitre plate (spots bigger than 1 mm in diameter are cut into smaller pieces).
2. Pipette off any liquid in wells using a multi-way pipette. (For this and all subsequent

Protocol 17 continued

pipetting steps use an eight-way multipipette and sterile disposable pipette tips with aerosol reduction filters as for PCR procedures.)

3. Add 100 μl of distilled water.
4. Remove liquid from wells.
5. Add 100 μl of distilled water.
6. Remove liquid from wells.
7. Add 60 μl destain solution (30 mM potassium ferricyanide; 100 mM sodium thiosulphate). Leave for 10 min.
8. Wash 100 μl of water.
9. Remove liquid from wells.
10. Add 120 μl acetonitrile. Gel pieces should shrink and appear white.
11. Remove liquid from wells.
12. Add 30 μl freshly prepared dithiothreitol solution (6.5 mM dithiothreitol in 50 mM ammonium bicarbonate pH 8.5). Leave for 60 min at room temperature.
13. Remove liquid from wells.
14. Add 60 μl acetonitrile. Shrink gel.
15. Remove liquid from wells.
16. Add 30 μl freshly prepared iodoacetamide solution (54.0 mM iodoacetamide in 50 mM ammonium bicarbonate pH 8.5). Leave for 30 min at room temperature.
17. Remove liquid from wells.
18. Add 60 μl acetonitrile. Shrink gel.
19. Remove liquid from wells.
20. Add 30 μl 50 mM ammonium bicarbonate pH 8.5.
21. Remove liquid from wells.
22. Add 60 μl acetonitrile. Shrink gel.
23. Remove liquid from wells.
24. Add 30 μl 50 mM ammonium bicarbonate pH 8.5.
25. Remove liquid from wells.
26. Add 100 μl acetonitrile. Shrink gel.
27. Remove liquid from wells.
28. Transfer 96-well plate to a rotary evaporator and dry gel pieces.
29. Leave to dry for 60 min or overnight.
30. Prepare enzyme just prior to use. For example, bovine trypsin, sequencing grade Boehringer Mannheim, Product No. 1418–475. (To prepare enzyme stock, add 1 ml

Protocol 17 continued

0.1% trifluoroacetic acid (TFA) to 100 mg of solid enzyme. Dilute 100 ml of this stock in a further 1 ml of 50 mM ammonium bicarbonate pH 8.5 to produce an enzyme concentration of 10 ng ml^{-1}.)

31 Add 10 μl enzyme solution to each well containing dry gel pieces.

32 Wait 30 min. Then add 20 μl 50 mM ammonium bicarbonate pH 8.5.

33 Leave overnight at room temperature or at least 3–6 h before sampling supernatant for MALDI analysis. Plates can be sealed with broad adhesive tape to reduce evaporation during this step.

34 An additional 20 μl of 50 mM ammonium bicarbonate pH 8.5 may be required if no liquid is present at the bottom of the wells.

Notes: Although lengthy, the procedure is simple and easily becomes routine or adapted for robotics. Maintain a cover over samples at all times, except during pipetting. Otherwise, sterile air is delivered continuously into the robotic space.
Many groups prefer the use of autolytic trypsin, as opposed to modified trypsin (Promega), because the digestion products can be employed for mass calibration and spectral analysis. If one employs routinely Trypsin and Glu-C, one can achieve excellent statistical confidence during protein characterization for all proteins able to be digested by both enzymes.
Quiet often protein samples have undergone reduction and alkylation prior to gel electrophoresis. Nonetheless, it is recommended to conduct this process again prior to protein characterization. The more efficiently this process is conducted, the greater is the likelihood of efficient enzymatic digestion (and thus the more peptide fragments closer to predicted mass) and/or reliable release of cysteine residues during Edman microsequencing. Reduction and alkylation also helps prevent degradation of cysteine residues to similar products as serine during Edman chemistry. It is noteworthy that reduction is best conducted at temperatures above ambient. Thus, step 12 can be conducted respectively for 15 min at 60 °C or 5 min at 90 °C. Further variations on recommended procedures for maximizing reduction and alkylation have appeared in the literature (94, 95).
We also recommend the protocol of Castellano-Serra *et al.* for 'in gel' digestion (96).

Protocol 18

MALDI stage cleaning just prior to loading

Equipment and reagents

- Acetone
- Distilled water

Method

1. Scrub stage under warm tap-water with non-abrasive sponge and detergent.
2. Wash with acetone.

Protocol 18 continued

3 Rinse well with tap-water.
4 Rinse well with distilled water.
5 Rinse with acetone.
6 Dry with compressed air to remove moisture from within grooves on the target.

Protocol 19

MALDI stage loading

Equipment and reagents

- α-cyano-4-hydroxy cinnamic acid
- 95% Acetone
- 5% formic acid
- 2% TFA
- Matrix solution: α-cyano-4-hydroxy cinnamic acid, acetone, nitrocellulose, acetone/propan-2-ol (1:1 v/v), 0.1% TFA

Method

1 Prepare a concentrated solution of α-cyano-4-hydroxy cinnamic acid. (Rinse the solid in 4 volumes of acetone. Decant off initial solvent. Dissolve α-cyano-4-hydroxy cinnamic acid maximally at room temperature in 95% acetone/5% water. The aqueous content is designed to limit damage to the 'thin film' during target blow-off.)

2 Prepare matrix solution in the lid of a microcentrifuge tube by mixing: 75 μl of α-cyano-4-hydroxy cinnamic acid in acetone; 20 μl of nitrocellulose (small pieces derived from a membrane) at 10 mg ml^{-1} in acetone/propan-2-ol (1:1 v/v); and 5 μl of 0.1% TFA.

3 Spot freshly made matrix solution on to each target position required. Upon evaporation a thin layer ('film') of matrix will remain. (Use a 800 μm diameter dispensing pin, as used for robotic applications, to obtain reproducible matrix spots during manual target loading. Otherwise, spot to target using a fine micropipette tip.)

4 Carefully add 0.3 μl of 5% formic acid to each of the target positions going to be used.

5 Add 0.5 μl of sample solution directly into the acid already on the target, i.e. before the acid in step 4 evaporates. (Alternatively, the liquid remaining following 'in-gel' digestion (*Protocol 17*) is sampled directly in an equal volume of 5% formic acid and spotted on to the stainless steel target pre-coated with matrix, as in step 3.)

6 Allow the target to air dry under a cover, so as to avoid dust deposition.

7 Briefly wash the target with 2 μl of 2% TFA in water or 5% formic acid in water. Excess wash solution should be blown off the target using compressed air.

8 Load the target plate into the mass spectrometer.

(After the 'thin film' method of Jensen *et al.* (97).)

7.4 Electrospray ionization mass spectrometry

ESI MS was introduced in the late 1980s as an attractive option for on-line delivery of peptides destined for mass spectrometry (88–90). Previously, it had not been possible to achieve vaporization through heating and subsequent ionization by atom bombardment. This technical dilemma was solved through rapid transition from liquid to gas phase in an ionized state, or Colombation. When combined with micro- and nano-electrospray, very small volumes of analyte were able to yield spectra over an extended period, e.g. 1 μl per 40–60 min. Nano-electrospray delivery is accomplished by housing sample within a very fine, gold-coated capillary needle across which is placed a potential difference (voltage) with respect to the inlet of the mass spectrometer. Slight manual pressure from a syringe is used to initiate analyte delivery to the mass spectrometer interface. Thereafter, the sample flow rate is determined by the process of Colombation. During liquid chromatography (LC)/ESI, it is the LC component that determines the flow rate. Under the influence of the electric field, highly charged small droplets disperse electrostatically towards the mass spectrometer. By a process of field desorption and/or solvent evaporation, the charge per droplet size increases rapidly to a critical value, at which time the droplet explodes into naked ions carrying a variety of charges, depending upon the ability of these ions to accept charge (see *Figure 8*; after 98). Weak acid in the initial delivery solvent is employed so as to increase the likelihood of protonation. Desorption of peptide ions from droplets into the gas phase is aided by a countercurrent of warm nitrogen gas to help dry the residual solvent surrounding peptides. Both salts and detergents interfere with the above process and thus sample clean-up by reverse phase HPLC is recommended prior to mass analysis.

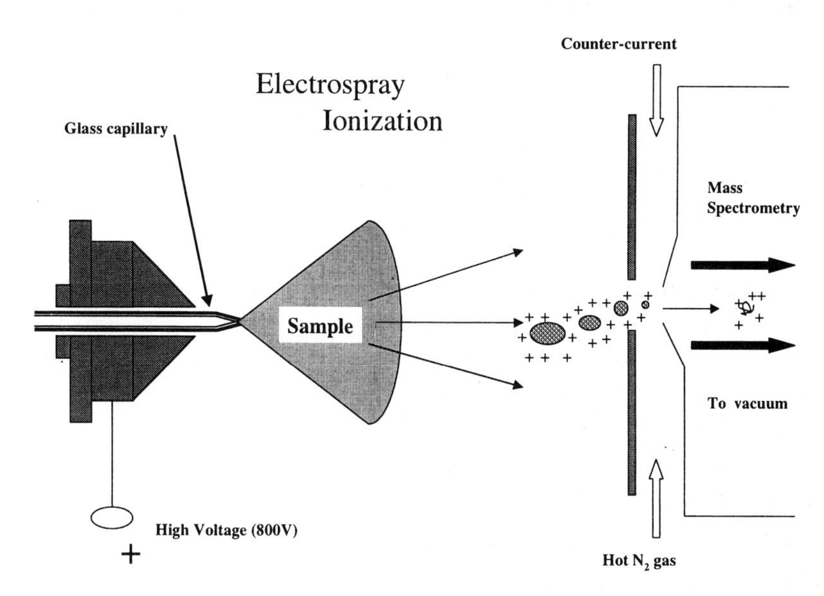

Figure 8 Principles underlying electrospray ionization, after ref. (98).

Following ESI, the protein sample can be analysed directly by mass spectrometry. Initially, the intact peptide masses are determined by virtue of the mass-to-charge (m/z) ratio observed for their respective molecular ions.

Tandem MS analysis (see below) may also be performed on individual precursor ions using either a triple quadrupole (*Figure 9*), an ion trap (*Figure 10*) and, more recently, a quadrupole TOF (*Figure 11*). These allow for 'sequence-tagging' or *de novo* sequence information to be extracted from individual percursor ions.

Tandem MS using a triple quadrupole mass spectrometer is based upon using Q1 to gate a selected precursor ion of interest before it is introduced to a

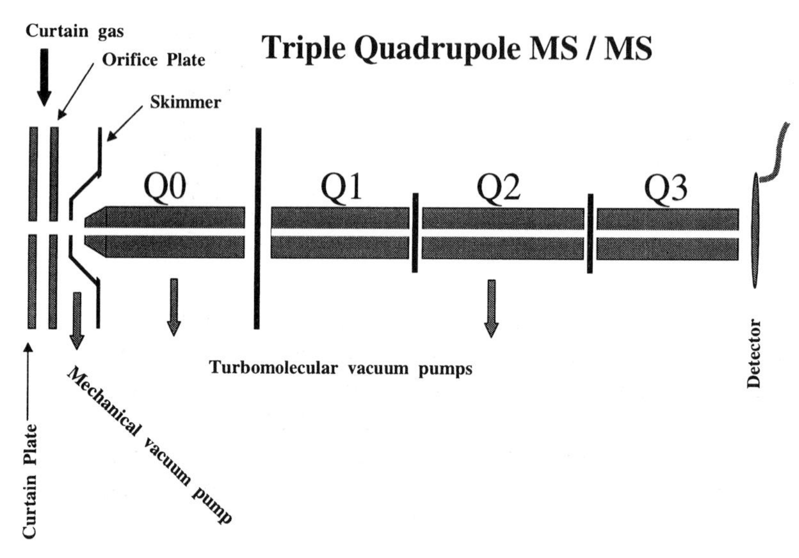

Figure 9 A schematic representation of triple quadrupole mass spectrometry.

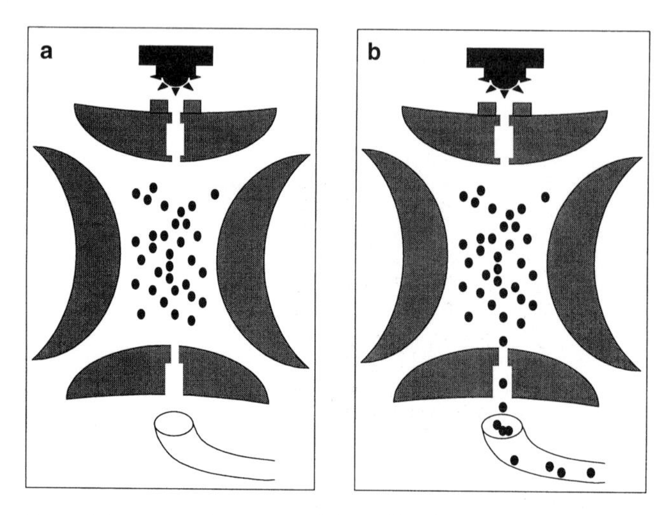

Figure 10 A schematic representation of ion trap mass spectrometry.

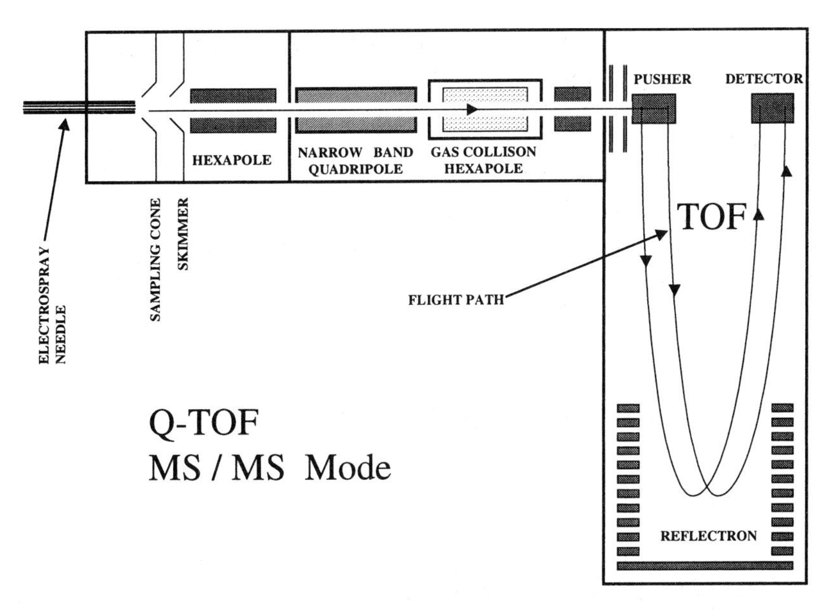

Figure 11 A schematic representation of quadrupole TOF mass spectrometry.

collision gas in Q2. Fragment ions produced by low energy collisions with argon gas can then be resolved in Q3 before reaching the detector (*Figure 9*). Q-TOF technology combines the ion handling capacity of a quadrupole instrument with advantages of added resolution and duty cycle provided by a TOF analyser.

As an add-on to liquid chromatography and electrospray ionization, the ion-trap mass spectrometer also provides a powerful and increasingly user-friendly environment for use in protein characterization. As the name would suggest, dual parabolic trapping potentials are exploited to confine ion species for subsequent analysis with respect to their respective mass/charge ratios by an

Nano-scale sample desalting and clean-up

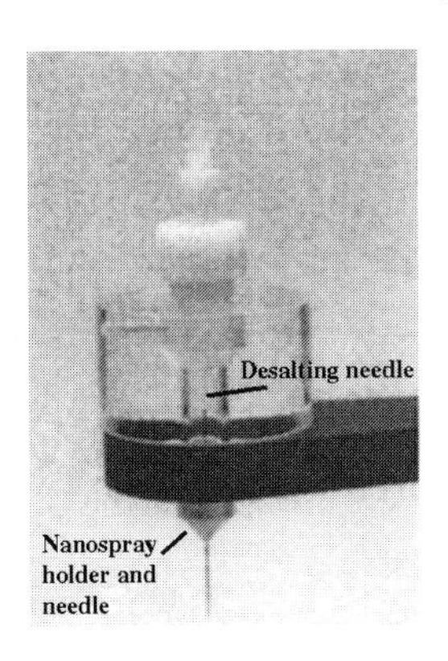

Figure 12 Nano-scale sample clean-up apparatus, showing capillary column and gold-plated nano-electrospray needle.

external detector (*Figure 10*). Ejection of ion species to the detector is achieved by linearly ramping the amplitude of the radiofrequency applied to an ion trap electrode so as to elute a particular ion series. The latter can be generated through collision-induced dissociation using a heavy gas, such as argon and xenon. The apparatus can be exploited for the generation of tandem mass spectra by conducting one mass selective operation after another by herding ions together under the influence of momentum-dissipating collisions with helium atoms. Unlike a triple quadrupole where parent ions are selected one after the other in an linear configuration, the ion trap employs pulsed extraction, so as to mass specifically accumulate ions over time and afford desirable sensitivity (99).

Protocol 20

Sample preparation and delivery to nano-electrospray mass spectrometry

Equipment and reagents

- 3-mercaptopropyl-trimethoxysilane
- 60% methanol/5% formic acid or 60% acetonitrile/5% formic acid
- Poros R2 resin (Perceptive Biosystems)
- P-97 Microcapillary Puller (Sutter Instruments)

Method

1. Borosilicate glass capillaries are pulled using a P-97 Microcapillary Puller using a two-step process. Jensen *et al.* (100) recommend the following procedure; however, a degree of trial and error may be necessary to optimize needle tip and flow rate to the mass spectrometer: heat; pull; velocity; time respectively for step 1 at 520, 100, 10, and 200 and for step 2 at 490, 160, 12, and 165.
2. Pulled needles are then gold coated in a Polaron SC 7610 sputter coater (Fisons Instruments). The gold coating lacks stability and sticks better to the glass surface is first treated with 3-mercaptopropyl-trimethoxysilane (101).
3. A non-gold-coated capillary tube is then mounted within a housing similar to that shown in *Figure 12* and is placed immediately above a gold-coated capillary to allow sample clean-up and desalting just prior to delivery of sample via nano-electrospray to the mass spectrometer. The non-gold-coated capillary contains a small quantity (5 μl) of Poros R2 resin (Perceptive Biosystems). See *Protocol 21* for preparation of this microcolumn and injection of sample on to capillary column.
4. Mount the above assembly on to the ion source of the mass spectrometer and apply voltage to the needle and sample delivery orifice of the mass spectrometer.
5. Elute the peptide mixture into the gold-coated capillary tube using 0.5–2 μl of 60% methanol/5% formic acid or 60% acetonitrile/5% formic acid.

Protocol 21

Capillary Poros R2 column preparation

Equipment and reagents

- Methanol
- 0.5% formic acid
- Poros R2 resin (Perceptive Biosystems)

Method

1. Poros R2 resin is obtained from PE Biosystems.
2. Sediment 20 μl of resin three to five times in 1.5 ml of methanol, each time removing only the larger beads for reuse.
3. Transfer 5 μl of resin to a non-gold-coated capillary needle.
4. Using a low speed centrifuge (200–5000 g) or 5–10 psi positive pressure applied to the top of the capillary sediment the resin to the tip of the pulled capillary needle and then break off the capillary tube tip to create a larger opening.
5. Rinse the column with 5 μl of methanol.
6. Equilibrate the column using 5 μl of 0.5% formic acid and centrifuge-away or blow through all remaining formic acid.
7. Add the peptide mixture. This may be an aliquot of the supernatant liquid from the digest (10 μl) or extracted peptides resuspended in 10 μl of 5% formic acid. Pass the peptide mixture through the column using either low speed centrifugation or a pressure driven device (*Figure 12*).
8. Wash the column twice with 5 μl of 0.5% formic acid.
9. Proceed with peptide elution as in step 5, *Protocol 20*. Peptides can be eluted into a gold-coated capillary for immediate analysis by nano-electrospray MS or a small microcentrifuge tube.
10. Discard the capillary column assembly. Do *not* reuse the columns, as peptides may contaminate subsequent samples.

After Jensen *et al.* (100) and adapted by MW.

7.5 Tandem mass spectrometry

Tandem mass spectrometry is based upon fragmentation of parent ions/precursor peptides. The success of this process can vary dramatically between peptides depending upon the charge state, the efficiency of collision-induced dissociation (CID), amino acid sequence, and the internal energy associated with the peptide ion. The nomenclature for fragmentation products was introduced by Roepstorff in 1984 (102) and subsequently modified (103) to that presented in *Figure 13*. Fragments must carry at least one charge. When this charge is retained on the amino-terminal fragment we refer to *a*, *b*, or *c* ions, or conversely *x*, *y*, or *z*

Table 4 Some common protein modifications

Modification	Mass	Modification	Mass
Acetylation	42	Hydroxylation	16
Amidation	−1	Methylation	14
Biotinylation	226	Myristoylation	210
Carbamylation	41	*N*-acetyl hexoseamines	203
Carboxylation	44	Oxidation	16
Deamidation	1	Palmitoylation	238
Deoxyhexose	146	Pentose	132
Formylation	28	Phosphorylation	80
O-GlcNac	203	Sialylation	291
Glucosylation	162	Sodium	22
Hexosamines	161	Sulphation	80

Table 5 Software available by internet and of relevance to protein characterisation

Server name	Address	Utility
ExPASy	http://expasy.proteome.org.au/	Variety of analysis packages and web links
PROWL	http://prowl.rockefeller.edu/	Variety of analysis packages and web links
PepSearch	http://www.mann.embl-heidelberg.de/ Services/PeptideSearch/ PeptideSearchIntro.html	Peptide mass fingerprinting and sequence tagging
MOWSE	http://www.seqnet.dl.ac.uk/mowse.html	Mass spectrometry analysis
Protein Prospector	http://prospector.ucsf.edu/	Mass spectrometry analysis
Propsearch	http://www.embl-heidelberg.de/prs.html	Amino acid analysis
GPMAW	http://130.225.147.138/gpmaw	Mass analysis software
EST databases	http://www.ncbi.nlm.nih.gov/dgEST http://sunny.ebi.ac.uk/ebi/EST/ http://ziggy.sanbi.ac.za/stack/stacksearch	When other approaches fail, check here for cDNA hits, particularly for human proteins
Mascot	http://www.matrixscience.com/cgi/ index.pl?page=/search_form_select.html	Mass spectrometry analysis
BasePeak	http://base-peak.wiley.com/msi/mssw.html	Shareware and web links
SEQUEST	http://thompsom.mbt.washington.edu/ sequest	Mass spectrometry analysis, not freely available
DeltaMass	http://www.abrf.oirg/ABRF/Research Committees/deltamass/deltamass.html	Mass modifications to proteins
KEGG EcoCYC WIT	http://www.genome.ad.jp/kegg/ http://ecocyc.pangeasystems.com/ ecocyc/ecocyc.html http://wit.mcs.anl.gov/WIT2	Metabolic pathways to increase confidence in a putative annotation; and web links
YOPD	http://www.proteome.com/	Functional annotations for *Saccharomyces cerevisiae* and *Caernorrhabditis elegans*

ions for fragments carrying charge on their carboxy-terminus. An additional proton derived from the precursor peptide may also be associated with *c* and *y* series ions. CID can produce ions showing the loss of either 17 Da for ammonia or 18 Da for water. The minimal internal fragment derived from either *a* or *y* cleavage event produces what is termed an immonium ion with characteristic mass for each amino acid residue. Amino acid incidence, residue and immonium ion mass appears in *Table 2*. Unfortunately, the masses cited in *Table 2* can vary due to any number of co- and post-translational modifications and also due to modifications of non-biological origin, some of the more commonly encountered mass discrepancies are shown in *Table 4*. *Figure 14* shows an example of a sequence tag obtained by tandem mass spectrometry.

Increasingly, mass spectrometry and protein characterization in general is supported by a number of bioinformatic tools accessible by the internet. *Table 5*

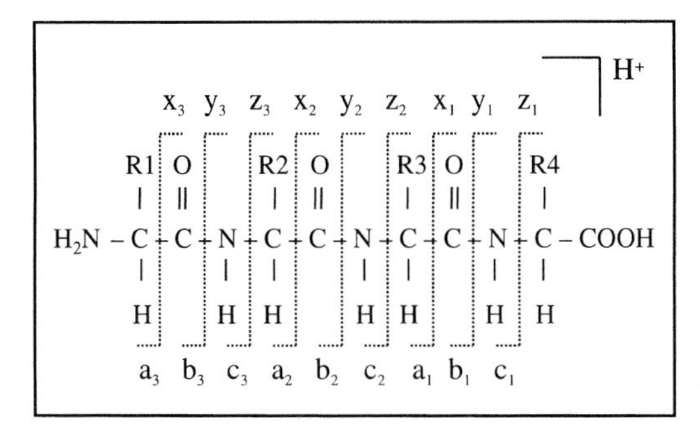

Figure 13 Nomenclature employed for ion-series generated during peptide fragmentation, after refs (102, 103).

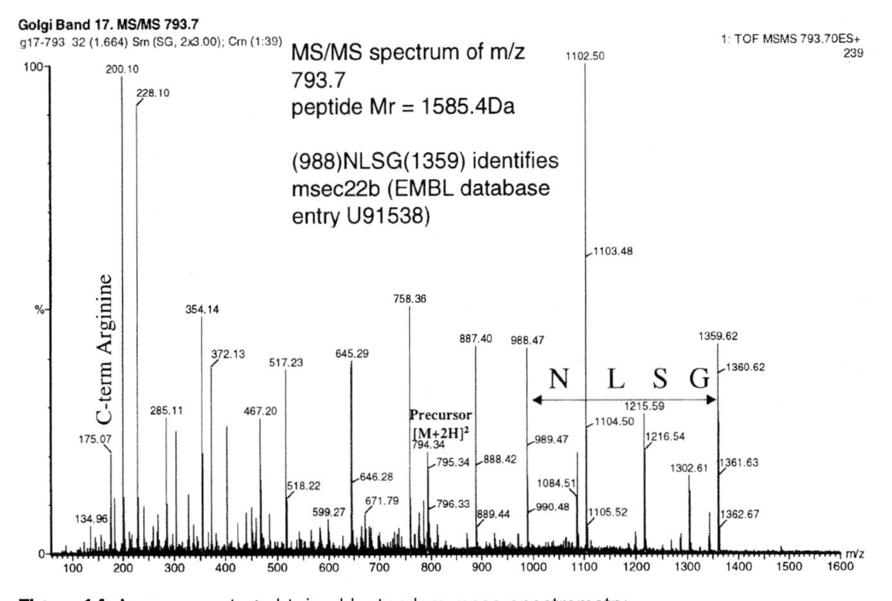

Figure 14 A sequence tag obtained by tandem mass spectrometry.

is a summary of some of the more useful web servers currently available in this sector.

Protocol 22

Methyl esterification and ^{18}O labelling of peptides

Equipment and reagents

- Thionyl chloride solution 1% (v/v) in anhydrous methanol
- 30% (v/v) acetonitrile
- 1% (v/v) methanol

Method

1. Following protein digestion, 1–2 µl of sample is placed in a microcentrifuge tube and dried using a rotary evaporator. Both the sample and the reaction vessels for this procedure must be devoid of water and are best dried in a heating oven at 110 °C prior to use.
2. Add 3–4 volumes of thionyl chloride solution 1% (v/v) in anhydrous methanol.
3. Seal the microcentrifuge tube and heat at 50 °C for 30 min in a heating block (NB: *not* a water bath!).
4. Dry the sample in a rotary evaporator.
5. Resuspend in 3 µl of 30% (v/v) acetonitrile; 1% (v/v) methanol.
6. Place a sample droplet on to MALDI stage (*Protocol 19*).
7. Comparison of spectra produced by tandem mass spectrometry (or post source decay spectra with masses displaced by 14 Da) with and without methyl esterification can permit amino acid sequenced to be non-ambiguously annotated (*Figure 15*).

Note: Alternatively, *y*-series ions can also be deciphered from other fragmentation ions as doublets separated by 2 Da and resulting from the use of ^{18}O-labelled water. In the latter case, the endoproteinase digest is performed with a 1:1 ratio of normal water to labelled water and both parent ions are selected simultaneously for subsequent fragmentation during MS/MS (see *Figure 16*).
(After Pappin (104), plus see notes on spectral interpretation (105).)

7.6 Analysis of post-translational modifications

Increasingly, mass spectrometry is being used to detail the nature of post-translational modifications and the location of these modifications within peptide sequences. Where common adducts are found in association with a particular amino acid residue, then it becomes a matter of recognizing the relevant mass discrepancy within a tandem mass spectrum (*Table 4*). This process must be conducted before and after enzymatic or chemical release of oligosaccharides, for example, and the subsequent detection of site-specific loss (106, 107). However, a more eloquent approach for the detection of known post-translational

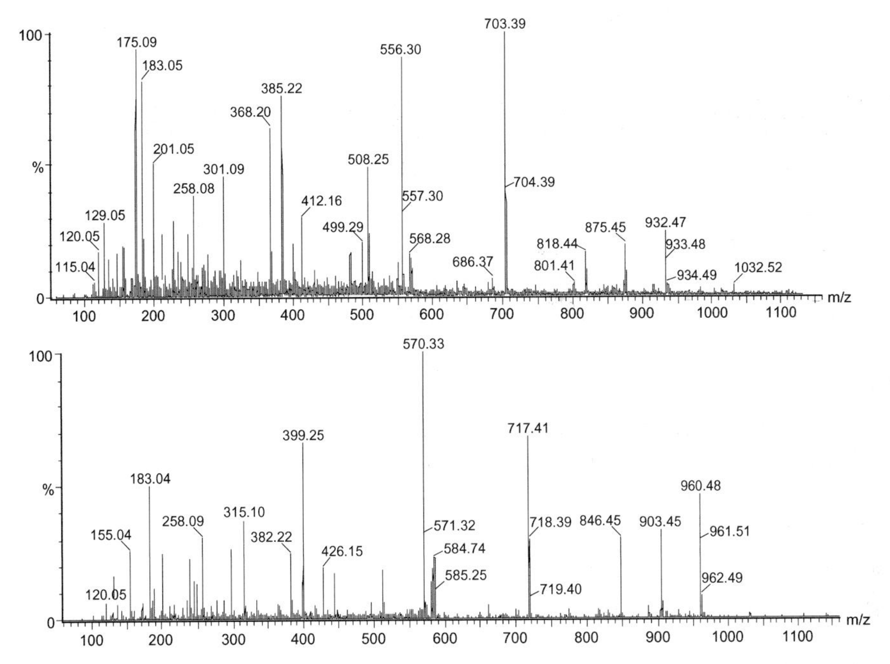

Figure 15 Comparison of MS/MS spectra for peptide sequence QAGGDFGNPLR before (top) and after methylation (bottom). The peptide molecular mass has increased by 28 Da (Doubly charged precursor ion originally m/z of 566.7 has shifted to m/z of 580.7). This indicates the presence of two acidic groups within the peptide, one is the C-terminus itself and the other is an aspartic acid residue at position 5.

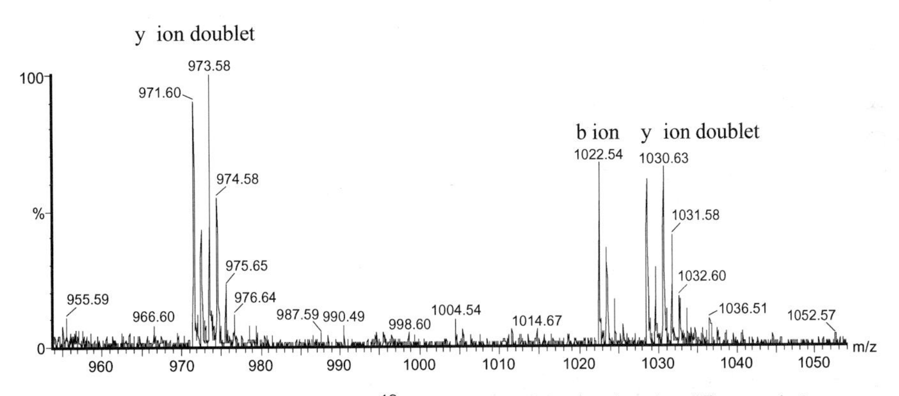

Figure 16 MS/MS spectrum of a 50% ^{18}O-labelled peptide showing the difference between *y* and *b* series ions.

modifications of interest is to undertake precursor ion scanning for the characterization of unseparated peptide mixtures (108, 109). Precursor ions (for example, m/z 79 for PO_3^-, as shown in *Figure 17*) are scanned in negative ion mode using nano-ESI and a triple quadrupole mass spectrometer. One then reverts to positive ion mode, the parent ion selected and fragmented by CID. A further example of an N-linked sugar is seen in *Figures 18* and *19*, whereby

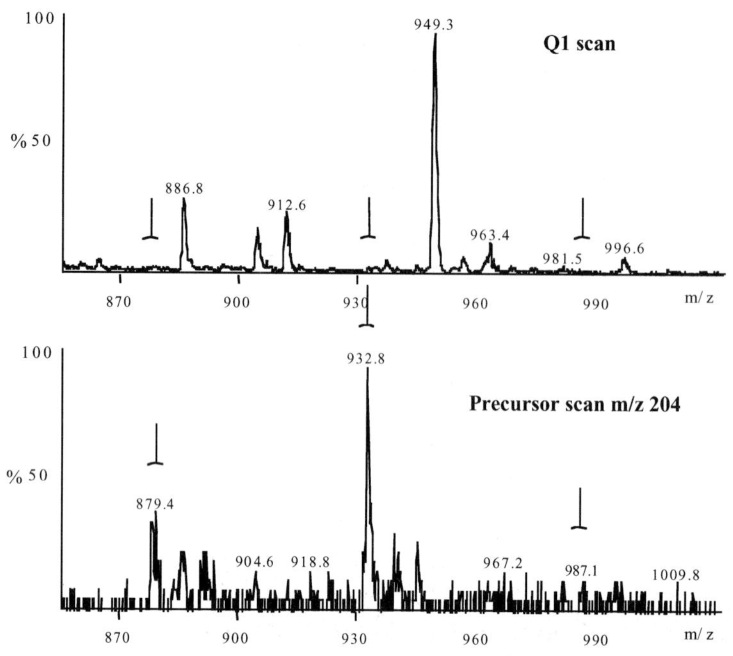

Figure 17 A product ion spectrum obtained following precursor ion scanning at m/z of 79 and highlighting a phosphoserine residue.

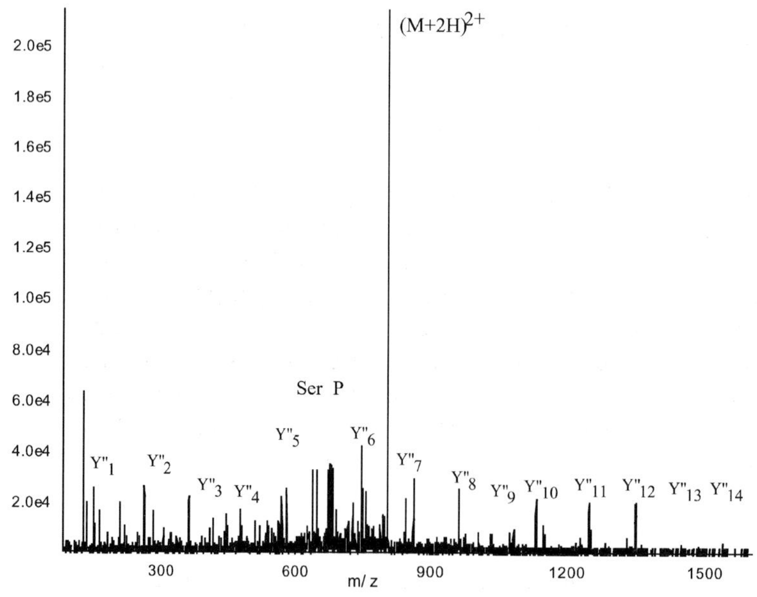

Figure 18 Comparison of Q1 and precursor m/z of 204 scanning for a tryptic digest of a glycoprotein. Note the absence of a parent ion at m/z of 933 in the Q1 spectrum.

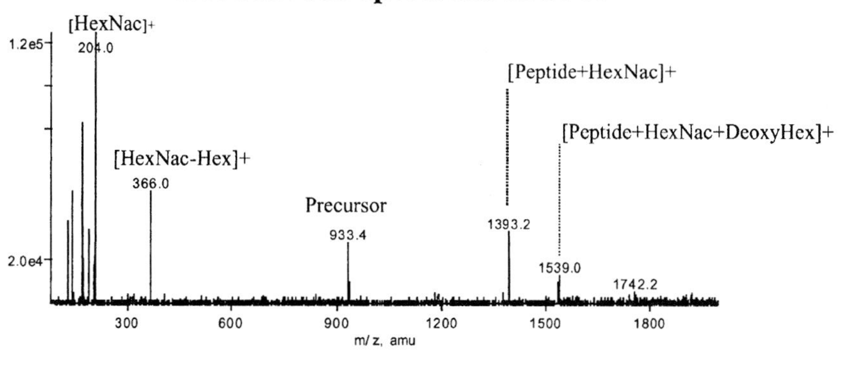

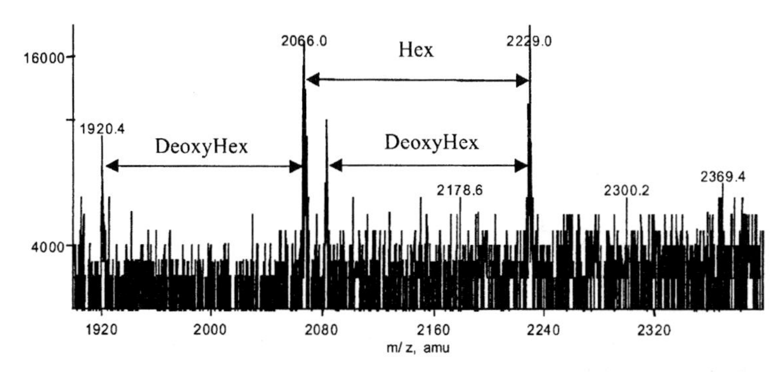

Figure 19 The product ion spectrum obtained in positive ion mode and following precursor ion scanning shown in Figure 16.

glycopeptides indistinguishable from within a peptide map (*Figure 18*) is subsequently detected and analysed following precursor ion selection (*Figure 19*). In this manner, only peptides bearing the appropriate modification are preselected for further analysis, for example, phosphorylation, sulphation and N- and O-linked glycans. This can represent a significant time saving when compared with manual scanning of individual MS/MS spectra of multiple peptides from within a large protein, not to mention the unlikelihood of detection if some of these peptides are among the least abundant during PMF. Neutral loss in positive ion mode is an equally attractive proposition for the detection of specific post-translational modifications. Preferential enrichment of a particular class of proteins may be conducted in advance by affinity chromatography, using metal chelating agents in the case of phosphoproteins, for example (108). More traditional approaches to the detection and characterization of post-translational modifications have been reviewed elsewhere (26,106, 107, 110, 111).

References

1. Wasinger, V. C., Cordwell, S. J., Cerpa-Poljak, A., Yan, J. X., Gooley, A. A., Wilkins, M. R., Duncan, M. W., Harris, R., Williams, K. L., and Humphery-Smith, I. (1995). *Electrophoresis*, **16**, 1090.

2. Kozian, D. H. and Kirschbaum, B. J. (1999). *Tibtech,* **17**, 73.
3. Wilm, M., Shevchenko, A., Houthaeve, T., Breit, S., Schweigerer, L., Fotsis, T., and Mann, M. (1996). *Nature,* **379,** 466.
4. Humphery-Smith, I. (1999). *Electrophoresis*, **20**, 653.
5. Jungblut, P., Thiede, B., Zimny-Arndt, U., Müller, E. C., Scheler, C., Wittmann-Liebold, B., and Otto, A. (1996). *Electrophoresis*, **17**, 839.
6. Roepstorff, P. (1997). *Curr. Opin.Biotechnol.,* **8**, 6.
7. James, P. (1997). *Biochem. Biophys. Res. Comm.*, **231**, 1.
8. Humphery-Smith, I. and Blackstock, W. P. (1997). J. Protein Chem., 16, 537.
9. Humphery-Smith, I., Cordwell, S. J., and Blackstock, W. P. (1997). *Electrophoresis*, **18**, 1217.
10. Wilkins, M. R., Williams, K. L., Appel, R. D., and Hochstrasser, D. F. (ed.). (1997). *Proteome research: New frontiers in functional genomics*, p. 243. Springer-Verlag, Berlin.
11. Lamond, A. I. and Mann, M. (1997). *Trends Cell Biol.*, **7**, 139.
12. Dainese, P., Staudenmann, W., Quadroni, M., Korostensky, C., Gonnet, G., Kertesz, M., and James, P. (1997). *Electrophoresis*, **18**, 432.
13. Yates, J. R. (1998). J. Mass Spectrometry, 33, 1.
14. Küster, B. and Mann, M. (1998). *Curr. Opinion Struct. Biol.*, **8**, 393.
15. Haynes, P. A., Gygi, S. P., Figeys, D., and Aebersold, R. (1998). *Electrophoresis* 19, 1862–1871.
16. Klose, J. (1999). *Electrophoresis*, **20**, 643.
17. Williams, K. W. *Electrophoresis*, **20**, 678.
18. Quadroni, M. and James, P. (1999). *Electrophoresis*, **20**, 664.
19. Persidis, A. (1999). *Nature Biotechnol.*, **16**, 393.
20. Dove, A. (1999). *Nature Biotechnol.*, **17**, 233.
21. Blackstock, W.P and Weir, W. P. (1999). *Tibtech*, **17**, 121.
22. Wang, J. and Hewick, R. (1999). *DDT,* **4**, 129.
23. Harris, E. L.V. and Angal, S. (ed.). (1990). *Protein purificiation methods*, p 336. IRL Press, Oxford.
24. Kellner, R., Lottspeich, F., and Meyer, H. E. (ed.). (1994). *Microcharacterisation of proteins*, p267. VCH, Weinheim.
25. Creighton, T. E. (ed.). (1997). *Protein structure. A practical approach*, p. 383. IRL Press, Oxford.
26. Shively, J. E. (1998). *Structural analysis of proteins*, p.100. In *Bioorganic Chemistry. Peptides and proteins* (ed. Hecht, S. M.), p. 532. Oxford University Press, Oxford.
27. Link, A. J. (ed.). (1999). *2-D proteome analysis protocols*, p. 601. Humana Press, New Jersey.
28. Howard, G. C. and Brown, W. E. (ed.). (1999). Basic methods in protein chemistry. CRC, Florida, In Press.
29. Karaoglu, H. and Humphery-Smith, I. (2000). In *Protein and Peptide Analysis: Advances in the Use of Mass Spectrometry.* (ed. Chapman, J. R.), In Press. Humana Press, New Jersey.
30. Patterson, S. D. and Aebersold, R. (1995). *Electrophoresis* **16,** 1791.
31. Andrews, A. T. (1986). *Electrophoresis: Theory, Techniques, and Biochemical and Clinical Applications.* (2nd edn), p. 452. Oxford Science Publications, Oxford.
32. Dunbar, B. S. (1987). *Two-Dimensional Electrophoresis and Immunological Techniques*, p. 372. Plenum Press, New York.
33. Dunn, M. J. (1987). *Adv. Electrophoresis* **1**, 1.
34. Hames, B. D. (ed.). (1998). *Gel Electrophoresis of Proteins: A Practical Approach.* (3rd edn), p. 352. IRL Press, Oxford.
35. Righetti, P. G. (1990). *Immobilized pH Gradients: Theory and Methodology*, p. 396. Elsevier Science Publications, Amsterdam.
36. Anderson, L. (1991)*Two-Dimensional Electrophoresis: Operation of the ISO-DALT System.* (2nd edn), p. 200. Large Scale Biology Press, Rockville.

37. Harrington, M. G. (1991). p 98. In *Methods: a companion to methods in enzymology.* Academic Press, London.
38. Dunn, M. J. (1993). *Gel Electrophoresis of Proteins*, p. 176. Bios Scientific Publishers, Oxford.
39. Dunn, M. J. (1997). *Methods Mol. Biol.,* **64**, 25.
40. Berkelman, T. and Stenstedt, T. (1998). *2-D Electrophoresis using immobilized pH gradients. Principles and methods*, p. 50. Amersham Pharmacia Biotech Inc., New Jersey.
41. Rabilloud, T. (ed.). (2000). *2D gel electrophoresis and detection methods.* P248, Springer-Verlag, Berlin.
42. Voris, B. P. and Young, D. A. (1980). *Anal. Biochem.* **104,** 478.
43. Young, D. A. (1984). *Clin. Chem.* **30,** 2104.
44. Klose, J. and Zeindl, E. (1984). *Clin. Chem.* **30,** 2014.
45. Patton, W. F., Pluskal, M. G., Skea, W. M., Buecker, J. L., Lopez, M. F., Zimmermann, R., Belanger, L. M., and Hatch, P. D. (1990). *Biotechniques* **8**, 518.
46. Levenson, R. M., Anderson, G. M., Cohn J. A., and Blackshear, P. J. (1990). *Electrophoresis*, **11,** 269.
47. Wasinger, V. C., Bjellqvist, B. J., and Humphery-Smith, I. (1997). *Electrophoresis,* **18,** 1373.
48. Neubauer, G., King, A., Rappsilber, J., Calvio, C., Watson, W., Ajuh, P., Sleeman, J., Angus Lamond, A., and Mann, M. (1998). *Nature Genetics* **20,** 46.
49. Grant, P. A., Schieltz, D., Pray-Grant, M. G., Yates, J. R. III and Workman, J. L. (1998). *Mol. Cell* **2,** 863.
50. Cordwell, S. J., Basseal, D. J., Bjellqvist, B. J., Shaw, D. C., and Humphery-Smith, I. (1997). *Electrophoresis* **18,** 1393.
51. Opiteck, G. J., Ramirez, S. M., Jorgenson, J. W., and Moseley, M. A. (1998). *Anal. Biochem.* **258,** 349.
52. Aebersold, R. A. (1991). *Adv. Electrophoresis* **4**, 81.
53. Baldo, B. A., (1994). *Adv. Electrophoresis* **7**, 407.
54. Edman, P. (1950). *Acta Chem. Scand.* **4**, 283.
55. Edman, P., Begg, G. (1967). *Eur. J. Biochem.* **1**, 80.
56. Laursen, R. A. (1971). *Eur. J. Biochem.* **20**, 89.
57. Hewick, R. M., Hunkapiller, M. W., Hood, L. E., Dreyer, W. J., *(1981). J. Biol. Chem.* **256**, 7990
58. Wasinger, V. C. and Humphery-Smith, I. (1998). *FEMS Microbiol. Lett.* **169**, 375.
59. Findlay, J. B.C. and Geisow, M. J. (eds.). (1989). *Protein sequencing. A practical approach*, p 199, IRL Press, Oxford.
60. Amo, M. and Tsugita, A. (1999). N-terminal amino acid sequencing of 2-DE spots, p 461. In *2-D proteome analysis protocols.* (ed. Link, A. J.). Humana Press, New Jersey.
61. Lottspeich, F., Houthaeve, T., and Kellner, R. (1994), In *Microcharacterisation of proteins*, (eds. Kellner, R., Lottspeich, F., and Meyer, H. E.), p 117. VCH, Weinheim.
62. Beynon, R. J., Bond, J. S. (1989). *Proteolytic enzymes: a practical approach.* P 259, IRL Press, Oxford.
63. Simpson, R. J., Nice, E. C. (1984). *Biochem. Int.* **8**, 787.
64. Walsh, M. J., McDougall, J., Wittmann-Liebold, B., *Biochemistry* 1988, *27*, 6867–6876.
65. Matsudaira, P. (1987). *J. Biol. Chem.* **262**, 10035.
66. Sanchez, J.-C., Ravier, F., Pasquali, C., Frutiger, S., Paquet, N., Bjellqvist, B., Hochstrasser, D. F., and Hughes, G. J. (1992). *Electrophoresis* **13**, 715.
67. Rasmussen, H. H., Van Damme, J., Puype, M., Gesser, G., Celis, J. E., and Vandekerckhove, J., (1991). *Electrophoresis* **12**, 873.
68. Cordwell, S. J. and Humphery-Smith, I. (1997). *Electrophoresis* **18**, 1410.
69. Kellner, R., Meyer, H. E., and Lottspeich, F. (1994), In *Microcharacterisation of proteins*, (eds. Kellner, R., Lottspeich, F., and Meyer, H. E.), p 93. VCH, Weinheim.

70. Wilkins, M. R., Yan, J. X., and Gooley, A. A. (1999). *2-DE spot amino acid analysis with 9-fluorenylmethyl chloroformate*, p445. In *2-D proteome analysis protocols*. (ed. Link, A. J.). Humana Press, New Jersey.
71. Wirth, P. J. and Romano, A. (1995). *J. Chromatography* **698**, 123.
72. Tomlinson, A. J., Guzman, N. A.and Naylor, S. (1995). *J. Cap. Electrophresis* **2**, 247.
73. Figeys, D., Ducret, A., Yates, J. R., III and Aebersold, R. (1996). *Nature Biotechnol.* **14**, 579.
74. Guzman, N. A., Park, S. A.S., Schaufelberger, D., Hernandez, L., Paez, X., Rada, P., Tomlinson, A. J., and Naylor, S. (1997). *J. Chromatography B*, 697, 37.
75. Heegaard, N., Nilsson, S., and Guzman, N. (1998). *J. Chromatograhy B*, **715**, 29.
76. Rosenfeld, J., Capdeveille, J., Guillemot, J. C., and Ferrara, P. (1992). *Anal. Biochem.* **203**, 173.
77. Hellman, U., Wernstedt, C., Gonez, J., and Heldin, C.-H. (1995). *Anal. Biochem.* **224**, 451.
78. Otto, A., Thiede, B., Müller, E.-C., Scheler, C., Wittman-Liebold, B., and Jungblut, P. (1996). *Electrophoresis* **17**, 1643.
79. Rider, M. H., Puype, M., Van Damne, J., Gevaert, K., De Boeck, S., D'Alayer, J., Rasmussen, H. H., Celis, J. E., and Vandekerckhove, J. (1995). *Eur. J. Biochem.* **230**, 258.
80. Kristensen, D. B., Inamatsu, M., and Yoshizato, K. (1997). *Electrophoresis* **18**, 2078.
81. Hatt, P. D., Quadroni, M., Staudenmann and James, P. (1997). *Eur. J. Biochem.* **246**, 336.
82. McLafferty, F. W., Fridriksson, E. K., Horn, D. M., Lewis, M. A., and Zubarev, R. A. (1999). *Science* **284**, 1289.
83. Mann, M. and Wilm, M. (1994). *Anal. Chem.* **66**, 4390.
84. Eng, J. K., McCormack, A. L., and Yates, J. R., III. (1994). *J. Amer. Soc. Mass Spectrom.* **5**, 976.
85. Wise, M. J., Littlejohn, T., and Humphery-Smith, I. (1997). *Electrophoresis* **18**, 1399.
86. Cordwell, S. J., Wilkins, M. R., Cerpa-Poljak, A., Gooley, A. A., Duncan, M., Williams, K. L., and Humphery-Smith, I. (1995). *Electrophoresis* **16**, 438.
87. Cordwell, S. J., Wasinger, V. C., Cerpa-Poljak, A., Duncan, M., and Humphery-Smith, I. (1997). *J. Mass Spectrometry* **32**, 373.
88. Whitehouse, C. M., Dreyer, R. N., Yamashita, M., and Fenn, J. B. (1985). *Anal. Chem.* **57**, 675.
89. Meng, C. K., Mann, M., and Fenn, J. B. (1988). *Zeit. Physik D* **10**, 361.
90. Fenn, J. B., Mann, M., Meng, C. K., Wong, S. F., and Whitehouse, C. M. (1989). *Science* **246**, 64.
91. Karas, M. and Hillenkamp, F. (1989). *Anal. Chem.* **60**, 2299.
92. Strupat, K., Karas, M., and Hillenkamp, F. (1994). *Anal. Chem.* **66**, 464.
93. Shevchenko, A., Wilm, M., Vorm, O., and Mann, M. (1996). *Anal. Chem.* **68**, 850.
94. Mørtz, E., Vorm, O., Mann, M., and Roepstorff, P. (1994). *Biol. Mass Spectrom.* **23**, 249.
95. Sechi, S. and Chait, B. T. (1998). *Anal. Chem.* **70**, 5150.
96. Castellanos-Serra, L. R., Fernandez-Patron, C., Hardy, E., Santana, H., and Huerta, V. (1997). *J. Prot. Chem.* **16**, 415.
97. Jensen, O. N., Podtelejnikov, A. V., and Mann, M. (1997). *Anal. Chem.* **69**, 4741.
98. Mann, M. and Wilm. M. (1995). *TIBS* **20**, 219.
99. Marsh, R. E. (1997). *J. Mass Spectrometry* **32**, 351.
100. Jensen, O. N., Wilm, M., Shevchenko, A., and Mann, M. (1999). Peptide sequencing of 2-DE gel-isolated proteins by nano-electrospray tandem mass spectrometry, p571. In In *2-D proteome analysis protocols*. (ed. Link, A. J.). Humana Press, New Jersey.
101. Kriger, M. S., Cook, K. D., and Ramsey, R. S. (1995). *Anal. Chem.* **67**, 385.
102. Roepstorff, P. and Fohlman, J. (1984). *Biomed. Mass Spectrometry* **11**, 601.
103. Beimann, K. (1990). *Applications of tandem mass spectrometry to peptide and protein structure*, p 167. In *Biological Mass Spectrometry* (eds. Burlingame, A. L. and McCloskey, J. A.). Elsevier, Amsterdam.

104. Pappin, D. J.C., Rahman, D., Hansen, H. F., Barlet-Jones, M., Jeffery, W., and Bleasby, A. J. (1995). In *Mass spectrometry in the biological sciences*. (eds. Burlingame, A. L. and Carr, S. A.), p135. Humana Press, New Jersey.
105. Courchesne, P. L. and Patterson, S. D. (1999). In *2-D proteome analysis protocols*. (ed. Link, A. J.), p 487. Humana Press, New Jersey.
106. Gooley, A. A. and Packer, N. H. (1997). In *Proteome research: New frontiers in functional genomics* (eds. Wilkins, M. R., Williams, K. L., Appel, R. D., and Hochstrasser, D. F.), p 65. Springer-Verlag, Berlin.
107. Packer, N. H., Ball, M. S., and Devine, P. L. (1999). Glycoprotein detection of 2-D separated proteins, p341. In *2-D proteome analysis protocols*. (ed. Link, A. J.). Humana Press, New Jersey.
108. Betts, J. C., Blackstock, W. P., Ward, M. A., and Anderton, B. H. (1997). *J. Biol. Chem.* **272**, 12922.
109. Neubauer, G. and Mann, M. (1999). *Anal. Chem.* **71**, 235.
110. Gooley, A. A. and Williams, K. L. (1997). *Nature*, **385**, 557.
111. Meyer, H. E. (1994). Analysing post-translational protein modifications, p 131. In *Microcharacterisation of proteins* (eds. Kellner, R., Lottspeich, F., and Meyer, H. E.). VCH, Weinheim.

List of suppliers

Amersham Pharmacia Bio Tech
Pharmacia Biotech (Biochrom) Ltd., Unit 22, Cambridge Science Park, Milton Road, Cambridge CB4 0FJ, UK
Tel: 01223 423723
Fax: 01223 420164
URL: http://www.biochrom.co.uk
Pharmacia and Upjohn Ltd., Davy Avenue, Knowlhill, Milton Keynes, Buckinghamshire MK5 8PH, UK
Tel: 01908 661101
Fax: 01908 690091
URL: http://www.eu.pnu.com

Anderman and Co. Ltd., 145 London Road, Kingston-upon-Thames, Surrey KT2 6NH, UK
Tel: 0181 541 0035
Fax: 0181 541 0623

Beckman Coulter Inc
Beckman Coulter Inc., 4300 N Harbor Boulevard, PO Box 3100, Fullerton, CA 92834-3100, USA
Tel: 001 714 871 4848
Fax: 001 714 773 8283
URL: http://www.beckman.com
Beckman Coulter (UK) Ltd., Oakley Court, Kingsmead Business Park, London Road, High Wycombe, Buckinghamshire HP11 1JU, UK
Tel: 01494 441181
Fax: 01494 447558
URL: http://www.beckman.com

Becton Dickinson and Co.
Becton Dickinson and Co., 21 Between Towns Road, Cowley, Oxford OX4 3LY, UK
Tel: 01865 748844
Fax: 01865 781627
URL: http://www.bd.com
Becton Dickinson and Co., 1 Becton Drive, Franklin Lakes, NJ 07417-1883, USA
Tel: 001 201 847 6800
URL: http://www.bd.com

Bio 101 Inc.
Bio 101 Inc., c/o Anachem Ltd., Anachem House, 20 Charles Street, Luton, Bedfordshire LU2 0EB, UK.
Tel: 01582 456666
Fax: 01582 391768
URL: http://www.anachem.co.uk
Bio 101 Inc., PO Box 2284, La Jolla, CA 92038-2284, USA
Tel: 001 760 598 7299
Fax: 001 760 598 0116
URL: http://www.bio101.com

Biodesign Inc.
New York, USA

Bio-Rad Laboratories Ltd
Bio-Rad Laboratories Ltd., Bio-Rad House, Maylands Avenue, Hemel Hempstead, Hertfordshire HP2 7TD, UK
Tel: 0181 328 2000
Fax: 0181 328 2550
URL: http://www.bio-rad.com

Bio-Rad Laboratories Ltd., Division Headquarters, 1000 Alfred Noble Drive, Hercules, CA 94547, USA
Tel: 001 510 724 7000
Fax: 001 510 741 5817
URL: http://www.bio-rad.com

Corning Costar, Corporate Headquarters, 45 Nagog Park, Acton, MA 01720, USA
Tel: (800) 492 1110 Fax: (508) 536 2476

CP Instrument Co. Ltd., PO Box 22, Bishop Stortford, Hertfordshire CM23 3DX, UK
Tel: 01279 757711 Fax: 01279 755785
URL: http://www.cpinstrument.co.uk

CLONTECH Laboratories
1020 East Meadow Circle
Palo Alto, CA 94 303-4320, USA
Tel: 650 424 8222 Fax: 650 424 0133
URL: www.clontech.com

Dupont
Dupont (UK) Ltd., Industrial Products Division, Wedgwood Way, Stevenage, Hertfordshire SG1 4QN, UK
Tel: 01438 734000
Fax: 01438 734382
URL: http://www.dupont.com

Dupont Co. (Biotechnology Systems Division), PO Box 80024, Wilmington, DE 19880-002, USA
Tel: 001 302 774 1000
Fax: 001 302 774 7321
URL: http://www.dupont.com

Dynal, PO Box 158 Skøyen, N-0212 Oslo, Norway
Tel: +4722061000 Fax: +47 22 50 70 15
e-mail: dynal@dynal.no

Eastman Chemical Co., 100 North Eastman Road, PO Box 511, Kingsport, TN 37662-5075, USA
Tel: 001 423 229 2000
URL: http://www.eastman.com

Edge BioSystems, Gaithersberg, MD, USA

Epicentre Technologies, 1402 Emil Street, Madison, WI 53713, USA
Tel: 001 284 8474
Fax: 001 609 258 3088
URL: http://www.epicentre.com

Eurobio, 7, avenue de Scandinavie, 91953 Les Ulis Cedex B, France
Tel: 01 69 07 94 77
Fax: 01 69 07 95 34

Fisher Scientific

Fisher Scientific UK Ltd., Bishop Meadow Road, Loughborough, Leicestershire LE11 5RG, UK
Tel: 01509 231166
Fax: 01509 231893
URL: http://www.fisher.co.uk
Fisher Scientific, Fisher Research, 2761 Walnut Avenue, Tustin, CA 92780, USA
Tel: 001 714 669 4600
Fax: 001 714 669 1613
URL: http://www.fishersci.com

Fluka
Fluka, PO Box 2060, Milwaukee, WI 53201, USA
Tel: 001 414 273 5013
Fax: 001 414 2734979
URL: http://www.sigma-aldrich.com
Fluka Chemical Co. Ltd., PO Box 260, CH-9471, Buchs, Switzerland
Tel: 0041 81 745 2828
Fax: 0041 81 756 5449
URL: http://www.sigma-aldrich.com

FMC Bioproducts, 191 Thomaston Street, Rockland, ME 04841, USA
Tel: 207 594 3495 Fax: 207 594 3491
e-mail: BIOTECHSERV@FMC.COM

Hamamatsu, Hamamatsu City, Japan

Hybaid
Hybaid Ltd., Action Court, Ashford Road, Ashford, Middlesex TW15 1XB, UK
Tel: 01784 425000 Fax: 01784 248085
URL: http://www.hybaid.com
Hybaid US, 8 East Forge Parkway, Franklin, MA 02038, USA
Tel: 001 508 541 6918
Fax: 001 508 541 3041
URL: http://www.hybaid.com

HyClone Laboratories, 1725 South HyClone Road, Logan, UT 84321, USA
Tel: 001 435 753 4584
Fax: 001 435 753 4589
URL: http://www.hyclone.com

Invitrogen
Invitrogen BV, PO Box 2312, 9704 CH Groningen, The Netherlands
Tel: 00800 5345 5345
Fax: 00800 7890 7890
URL: http://www.invitrogen.com
Invitrogen Corp., 1600 Faraday Avenue, Carlsbad, CA 92008, USA
Tel: 001 760 603 7200
Fax: 001 760 603 7201
URL: http://www.invitrogen.com

Life Technologies
Life Technologies Ltd., PO Box 35, Free Fountain Drive, Incsinnan Business Park, Paisley PA4 9RF, UK.
Tel: 0800 269210 Fax: 0800 838380
URL: http://www.lifetech.com
Life Technologies Inc., 9800 Medical Center Drive, Rockville, MD 20850, USA
Tel: 001 301 610 8000
URL: http://www.lifetech.com

Luigs and Neumann, Ratingen, Germany

MedProbe, MedProbe AS, PO Box 2640, St Hanshaugen
N-0131 Oslo, Norway

Merck Sharp & Dohme
Merck Sharp & Dohme Research Laboratories, Neuroscience Research Centre, Terlings Park, Harlow, Essex CM20 2QR, UK
URL: http://www.msd-nrc.co.uk
MSD Sharp and Dohme GmbH, Lindenplatz 1, D-85540, Haar, Germany
URL: http://www.msd-deutschland.com

Millipore
Millipore (UK) Ltd., The Boulevard, Blackmoor Lane, Watford, Hertfordshire WD1 8YW, UK
Tel: 01923 816375
Fax: 01923 818297
URL: http://www.millipore.com/local/UK.htm
Millipore Corp., 80 Ashby Road, Bedford, MA 01730, USA
Tel: 001 800 645 5476
Fax: 001 800 645 5439
URL: http://www.millipore.com

MJ Instruments, Watertown MA, USA

New England Biolabs, 32 Tozer Road, Beverley, MA 01915-5510, USA
Tel: 001 978 927 5054

Nikon
Nikon Corp., Fuji Building, 2–3, 3-chome, Marunouchi, Chiyoda-ku, Tokyo 100, Japan
Tel: 00813 3214 5311
Fax: 00813 3201 5856
URL: http://www.nikon.co.jp/main/index_e.htm
Nikon Inc., 1300 Walt Whitman Road, Melville, NY 11747-3064, USA
Tel: 001 516 547 4200
Fax: 001 516 547 0299
URL: http://www.nikonusa.com

Nycomed
Nycomed Amersham plc, Amersham Place, Little Chalfont, Buckinghamshire HP7 9NA, UK
Tel: 01494 544000 Fax: 01494 542266
URL: http://www.amersham.co.uk

Nycomed Amersham, 101 Carnegie Center, Princeton, NJ 08540, USA
Tel: 001 609 514 6000
URL: http://www.amersham.co.uk

Perkin Elmer Ltd., Post Office Lane, Beaconsfield, Buckinghamshire HP9 1QA, UK
Tel: 01494 676161
URL: http://www.perkin-elmer.com

Pharmacia (please see Amersham Pharmacia Bio Tech)

Promega
Promega UK Ltd., Delta House, Chilworth Research Centre, Southampton SO16 7NS, UK
Tel: 0800 378994
Fax: 0800 181037
URL: http://www.promega.com
Promega Corp., 2800 Woods Hollow Road, Madison, WI 53711-5399, USA
Tel: 001 608 274 4330
Fax: 001 608 277 2516
URL: http://www.promega.com

Qiagen
Qiagen UK Ltd., Boundary Court, Gatwick Road, Crawley, West Sussex RH10 2AX, UK
Tel: 01293 422911
Fax: 01293 422922
URL: http://www.qiagen.com
Qiagen Inc., 28159 Avenue Stanford, Valencia, CA 91355, USA
Tel: 001 800 426 8157
Fax: 001 800 718 2056
URL: http://www.qiagen.com

Robbins Scientific, Sunnyvale, CA, USA

Roche Diagnostics
Roche Diagnostics Ltd., Bell Lane, Lewes, East Sussex BN7 1LG, UK
Tel: 01273 484644
Fax: 01273 480266
URL: http://www.roche.com
Roche Diagnostics Corp., 9115 Hague Road, PO Box 50457, Indianapolis, IN 46256, USA
Tel: 001 317 845 2358
Fax: 001 317 576 2126
URL: http://www.roche.com
Roche Diagnostics GmbH, Sandhoferstrasse 116, 68305 Mannheim, Germany
Tel: 0049 621 759 4747
Fax: 0049 621 759 4002
URL: http://www.roche.com

Schleicher and Schuell Inc., Keene, NH 03431A, USA
Tel: 001 603 357 2398

Shandon Scientific Ltd., 93–96 Chadwick Road, Astmoor, Runcorn, Cheshire WA7 1PR, UK
Tel: 01928 566611
URL: http://www.shandon.com

Sigma-Aldrich
Sigma-Aldrich Co. Ltd., The Old Brickyard, New Road, Gillingham, Dorset XP8 4XT, UK
Tel: 01747 822211
Fax: 01747 823779
URL: http://www.sigma-aldrich.com
Sigma-Aldrich Co. Ltd., Fancy Road, Poole, Dorset BH12 4QH, UK
Tel: 01202 722114
Fax: 01202 715460
URL: http://www.sigma-aldrich.com

Sigma Chemical Co., PO Box 14508, St Louis, MO 63178, USA
Tel: 001 314 771 5765
Fax: 001 314 771 5757
URL: http://www.sigma-aldrich.com

Stratagene
Stratagene Europe, Gebouw California, Hogehilweg 15, 1101 CB Amsterdam Zuidoost, The Netherlands
Tel: 00800 9100 9100
URL: http://www.stratagene.com

Stratagene Inc., 11011 North Torrey Pines Road, La Jolla, CA 92037, USA
Tel: 001 858 535 5400
URL: http://www.stratagene.com

United States Biochemical, PO Box 22400, Cleveland, OH 44122, USA
Tel: 001 216 464 9277

Carl Zeiss Ltd, Welwyn Garden City, UK

Index